PEE WEE RUSSELL

• • • •

• • • •

PEE WEE RUSSELL

The Life of a Jazzman

ROBERT HILBERT

New York   Oxford
OXFORD UNIVERSITY PRESS
1993

Oxford University Press

Oxford New York Toronto
Delhi Bombay Calcutta Madras Karachi
Kuala Lumpur Singapore Hong Kong Tokyo
Nairobi Dar es Salaam Cape Town
Melbourne Auckland Madrid

and associated companies in
Berlin Ibadan

Copyright © 1993 by Robert Hilbert

Published by Oxford University Press, Inc.,
200 Madison Avenue, New York, New York 10016

Oxford is a registered trademark of Oxford University Press

Library of Congress Cataloging-in-Publication Data
Hilbert, Robert, 1939–
Pee Wee Russell : the life of a jazzman /
Robert Hilbert.
p. cm.
ISBN 0-19-507403-3
1. Russell, Pee Wee.
2. Jazz musicians—United States—Biography.
I. Title. ML419.R87H5 1993
788.6'2165'092—dc20 [B] 91–48155 MN

9 8 7 6 5 4 3 2 1

Printed in the United States of America
on acid-free paper

*To Betsy, Miriam and Marie Hilbert
with love and appreciation*

Preface

Pee Wee Russell has fascinated me since I first heard his chorus on Eddie Condon's recording of "Home Cooking" when I was 14 years old. His unique style was alternately praised and damned throughout his life. He was such an introvert that in some ways he could be said to have been an exhibitionist. It would be difficult to find a jazz figure whose style more reflected his personality. Pee Wee was what he played.

The story of his life was not easy to piece together. It covered most of the history of the development of jazz. He was the only musician to play and record both with Bix Beiderbecke and Thelonious Monk. "During the 1930s and '40s Russell was a familiar figure in jazz clubs," Leonard Feather wrote in his *New Encyclopedia of Jazz,* "playing with so great a variety of combos that complete documentation would be impossible." Indeed, this book does not attempt a complete chronology of his career but settles instead for the hope of some insight into his fascinating personality and thus of his music.

Since his death in 1969 Pee Wee's work has largely been ignored, except by a handful of astute critics and writers who continue to champion his music. This neglect is due to a variety of non-musical reasons. Jazz writers and critics, rather than dealing with a musician's uniqueness, often fit each artist into pre-existing categories, such as the New Orleans style, Chicago, swing, bop, cool, and so on. Because of the company he was usually forced to keep, Pee Wee was incorrectly pigeonholed as a dixielander when that term meant a musical reactionary. Actually he was one of the most revolutionary musicians in the history of the music. His intense improvisations only began to sound "right" to most critics during the last decade of his life. The groans and squeaks he elicited from his instrument were thought to be the result of not knowing the correct way to play. He did know the correct way, but he chose to play in his own way. In preparing this biography, I frequently encountered musicians who reacted in wonderment when I told them the name of my subject. Why not write about a *good* clarinetist, someone who knew how to play the instrument, like Benny Goodman or Artie Shaw (famous as leaders as well as instrumentalists)? Granted, Goodman and Shaw were superior technicians in the classical sense, but in my opinion Pee Wee outclassed both of them as a jazz improviser. One musician dismissed Pee Wee with a snort and said, "Can you imagine *him* playing the Mozart Clarinet Quintet with the Budapest String Quartet!" Exactly.

Russell spent most of his life in the shadow of others as a sideman. Unlike Goodman, Shaw and other "big names" who had full-time show business publicity mills grinding out copy and polishing their images as stars, Pee Wee never even had a manager let alone a press agent. He was not interested in show business; he thought that being a musician was enough. Despite that, he managed to attract quite a bit of press attention through the years, but his reclusive personality prohibited him from capitalizing on it. Alcoholism also colored and limited his life.

In more recent "Crow-Jim" times he had to bear the misfortune of having been a white jazz musician when only blacks need apply. As with so many of the other great white jazz contributors, this was particularly unjust in Pee Wee's case, because Pee Wee was color blind even as a youth living and playing in the segregated Southwest. Fortunately, we can still hear and appreciate his great work on the many recordings he made over 47 years. Although many of the other musicians on some of the recordings may sound dated and stale, Pee Wee's work remains as fresh and vital as ever. In fact, much of his work is more accessible today than it was when originally recorded. What was odd and irritating then sounds contemporary and invigorating now. Anyone who listens with unbiased ears will immediately appreciate his musical message. It has defied aging.

Piecing together the life and career of Pee Wee Russell involved the assistance of an army of enthusiasts and musicians. Without Dan Morgenstern's support, the project would never have been undertaken. Dan made the resources of the Institute of Jazz Studies at Rutgers University available to me, including notes compiled by George Hoefer, who had attempted an unfinished biography of the clarinetist in the 1960s. My indefatigable researchers, Lola Shropshire in Muskogee and Charles F. Rehkopf in St. Louis, dug up much invaluable information on Pee Wee's early life in those locales. Jerry and Wanda Simpson generously sent copies of the letters they had received from Pee Wee and his wife Mary. John McDonough was kind enough to provide me with the fruits of his research (including transcriptions of interviews with many musicians) gathered while writing his most informative essay on Pee Wee for the Time/Life boxed set of recordings. Hank Bredenberg generously provided his time and expertise in reviewing the manuscript.

Although no relatives of Pee Wee could be located, the assistance of his wife's nephews, Lee Goodman and his wife Estelle, and Sol Goodman and his wife Vera, was very helpful. Others who gave freely of their knowledge and time include:

Stephen Adamson; Tony Agostinelli; Henry Alberro; Ernest Anderson; Jay Anderson; Jeff Atterton; Doug Armstrong (Canada); George Avakian; Ray Avery; Bill Bacin; Dick M. Bakker (Holland); Whitney Balliett; Henry Behnke; John Bitter; Joe Boughton; Jack Bradley; Ruby Braff; Ace Brigode; Les Buchman; Paul Burgess; Mark Cantor; Hayden Carruth; Lou Carter; Dick Cary; David Chertok; John Chilton (England); Mac Clark; Derek Coller (England); Gerhard Conrad (Germany); Ken Crawford; Michael Cuscuna; Kenny Davern; Bob DeFlores; John Dengler; Frances A. Donelson; Frank Driggs; William Dunham, Wendell Echols; Don Ellis; Philip Evans; Tom Faber (Netherlands); John Featherstone (England); John Fell;

Bud Freeman; Milt Gabler; Jim Galloway (Canada); Russell
George; Joe Giorando; Vince Giordano; David Goldenberg; Don
Goldie; Robin Goodman; Jim Gordon; Bob Graf; Kenneth Gross;
Ed Grossmann; Tommy Gwaltney; Richard Hadlock; James R.
Hamilton; Ray Hatfield; Stanley Hester; Warren Hicks; Dick
Hill; Art Hodes; Paul J. Hoeffler (Canada); Franz Hoffman
(Germany); Steven Holzer; Peanuts Hucko; Tom Hustad; Rob-
ert Inman; Ed Jablonski; Sid Jacobs; Max Kaminsky; Harold
Kaye; Jack Keller; Peter Kennedy (England); Deane Kincade;
Larry Kiner; Shirley Klett; Wolfram Knauer (Germany); Gene
Kramer; Paul A. Larson; Steven Lasker; Ed Lawless; Floyd Le-
vin; Jack Litchfield (Canada); Giorgio Lombardi (Italy); Jim
Lowe (England); Earl Lyon; Tor Magnusson (Sweden); Irving
Manning: Tina McCarthy; Jim McHarg (Canada); Ann McKee;
Jimmy McPartland; Johnson McRee; Eugene Miller (Canada);
Keith Miller (Canada); Bill Miner; John Miner; Jack Mitchell
(Australia); Jamie Muller (Brazil); Joe Muranyi; John Nelson
(Canada); M. Nishiguchi (Japan); Hank O'Neal; Ted Ono; For-
rest "Fuzz" Pearson; Eugene Perkins; Don Peterson; Nat
Pierce; Ed Polic; Jon Pollack; Barrett Potter; Frank Powers; Jim
Prohaska; Al Pyle; Larry R. Quilligan; Dick Raichelson; Dolf
Rerink (Netherlands); Ed Reynolds; Evelyn Rich; Brian A. L.
Rust (England); Jack Sadler (Canada); Tom Saunders; Duncan
Schiedt; Dick Schindling; Rolf Schmidt (Germany); Yasuo
Segami (Japan); Manfred Selchow (Germany); Hans J. Schmidt;
Paul Sheatsley; Dick Shindeling; Joe Showler (Canada); Her-
bert L. Shultz; Dan Simms; Chuck Slate III; Jack Sohmer; Jess
Stacy; Alice Stevens; Jack Stine; Klaus Stratemann (Germany);
Dick Sudhalter; Mike Sutcliffe (Australia); Ralph Sutton; Bill
Thompson; Bob Thompson; Elmer Truch; Neville Twist (New
Zealand); Warren Vache, Sr.; Jerry Valburn; Ute Vladmir,
George Wein; Bob Weir (England); Bozy White; Bert Whyatt
(England); Jim Williams, Hal Willard; Satoshi Yuze (Japan).

Coral Gables, Florida   R. H.
1992

Contents

Foreword

DAN MORGENSTERN

Pee Wee Russell's approach to the clarinet—to jazz—was strictly his own. He was fearless, venturing into realms now called "space" long before that became fashionable, but he always landed safely. He knew his fundamentals.

Dick Wellstood (a gifted writer as well as a great pianist) described what he so rightly called the "miracle" of Pee Wee's playing as "that crabbed, chocked, knotted tangle of squawks with which he could create such woodsy freedom, such an enormously roomy private universe." Those sounds were indeed key colors in Pee Wee's tonal palette, but he could also coax contrastingly pure and gentle notes from his instrument.

I fell in love with the music of this extraordinary player long before I knew anything about him or had seen a photograph of his distinctive visage. This was during my first stage of serious interest in jazz. The setting was Denmark, right after World War II, and thus the focus was on records. I'd discovered a second-hand record and magazine store in Copenhagen that had a fascinating (if helter-skelter) stock at prices suited to a fifteen-year-old's budget. I would certainly have passed up a vintage HMV as by "Mound City Blue Blowers" if someone hadn't scrawled "C. Hawkins" on the label in white ink. Coleman Hawkins's name was already familiar to me, so I copped *Hello Lola* backed with *One Hour*. To be sure, Hawk delivered (these, after all, were landmark Hawkins solos), but it was the clarinet work that touched me deeply, especially Pee Wee's solo on *One*

Hour. In a phrase I had yet to learn, it "told a story." I played it over and over, and I love it still.

Then came the Rhythmakers things—pretty well represented on English Parlophone—and a taste of Pee Wee's tenor. Is his solo on *Anything for You* the first r&b tenor sax solo? By then (1932) his approach to the instrument had changed drastically from what we hear some five years earlier on *Honolulu Blues* or *Feelin' No Pain* (Molers version), which is so light and easy it makes one wonder if young Lester Young didn't hear it. But then, doesn't that *One Hour* solo bring Prez to mind, storywise?

Pee Wee was one of the great musical storytellers. Now we have his life story, between hard covers. He would no doubt have waved off the honor of a biography just as he waved off even the best-deserved applause, but he would have been pleased inside. Though he was often diffident to a fault, he knew his own worth. How else could he have stuck so stubbornly to his creative imperative? That took not only guts but conviction; he didn't have to play as he chose to do.

This timely book will correct some misconceptions. Since so much of his best-known work was done in relatively informal settings, Pee Wee is often perceived as an intuitive player, but as Bob Hilbert shows us, he was far from unschooled. Indeed, for some five years he was a member of that charmed circle of New York dance band, recording and radio studio musicians that also included Benny Goodman and the Dorsey Brothers. Pee Wee may not have been first call, like Benny and Jimmy, but he doubled on a bunch of reeds and did very well in what then was the world's most competitive professional environment for a non-classical musician. (This is confirmed in Bob Hilbert's Pee Wee Russell discography, published simultaneously with this volume.)

Light is also shed on Pee Wee's happy years with Louis Prima, largely ignored by the critical fraternity (excepting the often amazing Max Harrison). It was from musicians that I learned how much prime Pee Wee was on those Prima Brunswicks. Those two went their separate ways, but when I was about to see Prima in 1967, I told Pee Wee, who asked me to convey his warmest greetings. "I had some of the best times of my working life with that man," he said. "And he could play some trumpet too." Prima was thrilled to hear from Pee Wee. "How is that old so-and-so," he inquired, "is he taking care of himself?" Then he turned dead serious. "That man is a *genius*," he said.

Not all leaders understood Pee Wee so well. The Svengali-minded trombonist Marshall Brown organized a quartet for Pee Wee that had considerable musical potential, but after its very first public performance—at a press party thrown by George Wein—I knew it couldn't last long. When I congratulated Marshall after the set, he was brimful of himself. "Yes," he said. "It's the right combination. Pee Wee provides the feeling and I supply the musical intelligence." Boorishness aside, Pee Wee didn't particularly appreciate such Brown stratagems as the inclusion of pieces by John Coltrane and Ornette Coleman in the quartet's repertoire, for these were simply blues lines, picked for opportunistic reasons. Yet the quartet helped re-establish Pee Wee in the forefront of the music, where he'd always belonged.

After many years of observing Pee Wee in action in a wide variety of settings I got to know this by then very private man thanks to two of his great friends, Bobby Hackett and Jeff Atterton. Most important, Mary approved of me. That Pee Wee lived for almost two decades after his miraculous return from death's door was above all due to Mary's devotion and determination. They had both been legendary drinkers, but she cured herself and made him toe the line, with patience and humor. Incidentally, if we look at Pee Wee (and such other extraordinary grape consumers as, say, Eddie Condon and Wild Bill Davison) as "alcoholic," whatever that may mean, we simply miss the point. The environment in which these Prohibition-bred musicians grew up so predisposed them to become drinkers that it is neither funny nor tragic that they did. What is remarkable is that they nevertheless lasted so comparatively long. Surely that wasn't what George Avakian anticipated when he brought Pee Wee, Condon, et al. together in 1939 for a "last fling" at Chicago-style jazz. Since I am almost as old at this writing as Pee Wee was when he died I know that sixty-three ain't old age, yet I see him as a survivor. We were blessed to have him around that long.

It's not in fashion these days to cite Pee Wee Russell as one of jazz's truly great originals, though Gunther Schuller characteristically doesn't miss the boat—his chapter on Pee Wee in *The Swing Era* is required supplemental reading. But the records are evidence, here for new generations of listeners to discover. And while no one could ever again play like him, the spirit of Pee Wee's musical audacity and passion lives on in the gifted hands of Kenny Davern and in the surprising (and almost unknown outside Chicago) playing of Frank Chace. Both knew and loved Pee Wee, the man and his music.

In the end man and music are one, and this book, on which Bob Hilbert worked so hard and well, uncovering new facts and properly re-focusing others, will do much to cement Pee Wee Russell's rightful place in the pantheon of jazz. May it also bring new ears to his immortal music.

PEE WEE RUSSELL

• • • •

1 • • • •

A Bohemian
Existence

*J*azz, like the rest of the world, had chosen sides by 1944. The war, with its stark contest between good and evil, seemed simpler to fathom than the various factions in the jazz world. Any jazz fan asked to name his favorite clarinetist would instantly reveal to which of the hostile camps he owed his allegiance. A "mouldie-fygge," as they spelled it, might champion a New Orleans pioneer like Johnny Dodds or Jimmy Noone, or perhaps George Lewis. A "jitterbug" would probably choose one of the clarinet-playing bandleaders: Benny Goodman, Artie Shaw, Jimmy Dorsey or Woody Herman. A "hep cat" might balk at selecting any one, even with the emergence of Buddy De-Franco, Tony Scott and John LaPorta, because, after all, the clarinet itself was as "square" as the banjo. There was yet another group who thought of themselves as the keepers of the flame, who stirred the coals of the early hot jazz traditions begun by the Original Dixieland Jazz Band and carried to its highest expression by the great Bix Beiderbecke. For those hardy souls, there could be only one: Charles Ellsworth "Pee Wee" Russell, winner that year of the *Down Beat* poll for best clarinetist—in any style.

Pee Wee had held forth at Nick's, the center for dixieland jazz, off and on since 1937 while the bands changed around him. "I come with the lease," he said. At first, the Greenwich Village night spot had been a gathering place where the local bohemians—artists, writers, musicians, left-wing intellectuals—mingled with the "cafe society" set. But during World War II, Pee Wee's frequent appearances on radio programs transcribed for rebroadcast at military posts throughout the world had brought him new popularity. Soon, new fans in khaki elbowed their way to the bar.

For one private on leave, Ed Jablonski, simply hearing the music live at Nick's was not enough. He wanted to meet the clarinetist he had admired for so long. One afternoon, still fired up by the music he had heard the night before at Nick's, he set off for Pee Wee's apartment building, a decrepit wooden structure at 125 West 21st Street. There, he encountered Pee Wee's wife, Mary, on the second-floor landing and introduced himself, mentioning the name of mutual friends in Saginaw, Michigan, with whom Pee Wee and Mary had been corresponding. Jablonski later wrote to his Saginaw friends:

> She wore dark slacks, a shirt and a sweater and a coat. She had a package in each arm. About five four with black hair and not unattractive. Not beautiful either, but interesting. "Come up," she said, "and meet Pee Wee."

They walked up to the third floor, went through a maze of rooms and finally arrived at the apartment.

> She tossed the door open, and I followed her in. She put the groceries on the table at the left. She turned and quickly went into the other room and put up one of the shades. "Come on, Russell, get up. We have company."

The apartment consisted of a space approximately ten by fifteen feet, divided into two rooms by an arch.

The "kitchen," which I was in, had a table in it, a small gas stove opposite the door as you entered; to the right of that was a hideous cabinet of some kind in the corner. Immediately to your right was the sink. Pasted next to the mirror over it was a page from a magazine of *True Story* caliber with the words "I Was a Nagging Wife." When I closed the door there was a large Exit sign on it. Through the sign was a nail which had clothing dangling from it. The walls of the "kitchen" and the bedroom were plastered a cream color—once. Very few pictures—in the kitchen on the left wall over the table was a group shot of Pee Wee and some men singing. There was a calendar "Simpson Coffee Company." On that same wall was a door which led to the closet, which, when opened, can cause almost anything to happen.

The bedroom-living room consisted of a large dresser with a good sized mirror over it. That was marked "Reserved." The wall against which stood the dresser had large paneled wood doors, but I never saw them used at all. There was a chair in the corner. Then came the next wall which held two windows, between which stood a bureau. Over that was a poster showing Pee Wee on it, heralding "Eddie Condon's Jazz Concerts" at Town Hall. There were radiators under each window. On the one nearest the head of the bed, books were piled. I noted *Adventures of Archy and Mehitabel, I'm Thinking of My Darling,* and a western story.

Pee Wee was in bed. Sitting up, he looked quite dazed as Mary told him who I was. I shook his hand and muttered something about hearing him last night. His jet black hair sort of sprawled along both sides of his head, giving him a shaggy appearance. I noted something I had not seen the night before: two small scratches alongside his nose and on his cheek. Two deep scratches about two inches long graced his small chest. There was a small cyst just below the lobe of his right ear. His skin was pale. He seemed shaky as he lit a cigarette.

Mary coaxed him to get up. It was five o'clock. "Get up," she said, "and go to the store and get a chop for Ed."

When Pee Wee stands, his feet are slightly apart and his head hangs because he's somewhat unsteady. He usually holds onto something. He doesn't say much. And I was saying less.

Mary told Jablonski that Pee Wee hadn't come in till about 4:30:

> "We fought until about 5:15, read a while and went to bed. Pee Wee slept until five. I only just got up. Russell, you threw that ash tray at me."
>
> "Oh, no," Pee Wee laughed.
>
> "Yes, and I circled that date on the calendar. Wait—you were vicious too. Where's a pencil? There, I'll put it on: 'V' for vicious."

Pee Wee left to get the chop and a few other things. Mary took Jablonski with her through the maze of apartments to one that seemed identical to theirs. It was empty. They went in with Mary explaining that the apartment belonged to her friend, Lillian, with whom she ate every night. Within minutes, Lillian returned and the two women began preparing the meal.

> There were footsteps in the hall, and Lillian said, "We're having company. Pee Wee's coming." And it was. He poked his head in and smiled, and entered.
>
> Pee Wee wore a covert suit with a gray sweater pulled over his shirt. He held a cigarette in his hand. He seemed to lean, so I suggested he take the easy chair I occupied and I hopped into a seat at the foot of the bed. He fell heavily into the chair. He hadn't shaved. Mary came over and sat in his lap.

It was getting time for Pee Wee to get ready to go to work—he said the word with a slight shudder—so they returned to the Russells' apartment. Pee Wee shaved ("I gotta scrape," he told Jablonski), and examined his somewhat shabby tux. He owned a newer one, but a couple of months earlier he had lent it, a couple of shirts, and a black tie to one of the occupants of the building who presumably still had them.

"You know, Pee Wee," said Mary, "I saw him a few days ago. He still lives here on the second floor."

"Maybe I can get my clothes back," Pee Wee said. "I need them. I'll break into his room. I can't hit him, Ed, he's crippled—something wrong with his leg. I only let him have the suit to go to Washington to make seventy bucks. He's a musician. But it's been long enough. I saw him a whole month ago and he asked me if I was mad and I told him I was, but I couldn't do anything. I'll get the key to his room and get my stuff back. I'll break the door down if I have to." He dabbed at a shaving cut on his face with a towel which was draped around his slight shoulders.

"Pee Wee spoke to the landlord about the trombonist who had rooked him of a suit. The man said, 'Now, Pee Wee, why did you do that? He hasn't got a job.'"

Pee Wee said, "'I know, but he had a chance to make some money in Washington, so I let him have the suit. Let me have the key to his room. I can't hit the guy, they'd flush me away.'"

The landlord gave Pee Wee the keys, and he and Jablonski went up. "When we came to the door, Pee Wee gave me the keys," Jablonski wrote. "'You're steadier,' he said. The fifth key was the right one. We entered the room. It was smaller than Pee Wee's. When he found the light I could see that all it held was a bed, dresser, table and chair. The place was in a terrible mess. The bed was unmade and the sheet hadn't been washed in a month—on either side. Papers were scattered about the floor under and alongside the bed. The pile was about a foot deep, the latest *Down Beat* lay on top. A hat and tie hung on the chair. I began going through the dresser, the top of which was littered—one item being a badly soiled false shirt front. In the third drawer there was a brown shirt. Pee Wee took it. This is one, he said, and was appalled at the soiled condition. He began cursing the fellow roundly and then apologized to me for it. I told him I was better at it than he.

"Further search was fruitless and we decided to leave. As we turned to the door, Pee Wee said, "Look at this place. Only a moron or a degenerate could live in it!' We turned out the light, slammed the door and with Pee Wee leading the way, we descended into the catacombs, as he called it.

"Mary took the shirt and told Pee Wee she wanted half a dollar. He seemed a little reluctant but she finally got it. When he playfully embraced her, she said a strange thing: 'Why is it that of all people I despise you most?'

"We went out to the street and Pee Wee spoke. 'You shouldn't get us wrong,' he said. 'Take the landlord. He owns the building, this building and one further uptown where we might move to. Yet he comes here to play poker. He likes to be here, with people he knows. You're safe. You live the way you want.'

"We went to Seventh Avenue and got a cab. He continued there in this abstract way. 'You shouldn't be shocked, Ed, by the way we live.' Then he became silent but before very long we were there.

"'Stop at the liquor store,' Pee Wee directed, and we went in. He looked at the bottles like a little boy in a toy store. The expensive ones first. 'I'll be over there soon,' he said, pointing to the less costly bottles. He chose a bottle of rum and when the clerk asked if he wanted the operation performed, Pee Wee said: 'By all means, it must be done. I'm not capable of it myself.' The clerk then opened the bottle, stopped in the cork, and put it in a package. Pee Wee took it, tore the paper off, tested the cork to assure himself the bottle was open, then put it under his coat and we went out.

"'Must get some cigarettes.' So we went to a little place right next to Nick's. The woman got the pack and waited as Pee Wee laid a quarter on the counter, then a couple of pennies. We pushed out of the store.

Pee Wee's bohemian life-style, not very different from that of most of the artists, writers and musicians who lived in Greenwich Village at that time, was not what Jablonski had expected.

After all, other famous clarinetists like Benny Goodman or Artie Shaw did not live in decaying third floor walk-ups. But Pee Wee was venerated precisely because he was the antithesis of the smooth, slick, clarinet-playing band leader. He had refused to sell out to the big bands. His was the pure flame: hot, gritty, profane, real. No matter what physical or mental condition Russell was in, night after night he spun wondrous improvisations. No matter how disjointed his life, how scrambled his mind, how incomprehensible his speech, his music remained logical and authoritative, elegant and graceful, haughty and proud. Somehow, there was no dichotomy: the music was like steam escaping from the pressure cooker of his personality.

Pee Wee was born in the Maplewood section of St. Louis, Missouri, on March 27, 1906, to Charles Ellsworth Russell and Ella Ballard Russell. They were undoubtedly surprised when Edith became pregnant, since their marriage had been barren for twelve years. They were both 40 years old. Charles Ellsworth Russell, Jr., was their only child. Charles Russell, Sr., was born in August, 1863, and came to St. Louis from his native Ohio while still a young man. Shortly after arriving in St. Louis Russell, Sr. got a job as a tobacco broker. He married Ella Ballard, a young woman from Michigan, in 1894 and, by 1900, was living at 1367 Temple Place with his wife and members of her family. These were Ella's mother, Laura, Ella's brothers Milton C. Ballard and Will Ray Ballard, and a sister, Edith F. Ballard, born after the family moved from Washington, D.C., to Missouri. Will Ray was working as a dairyman and Milton and Edith were still in school when Pee Wee was born.

Pee Wee's father worked at a variety of jobs. Pee Wee always tried to portray his father as a successful businessman, but he seems to have never risen higher than the level of manager of small stores or clubs owned by others. His occupation is given as broker in the 1901 edition of *Gould's Directory of St. Louis*. He was still employed in that capacity in 1905. Around the time of Pee Wee's birth, his father got a job as a bartender at Joseph Marre and Sinopoulo Saloon in Maplewood. His next job was as the manager of a candy store. Beginning 1909, his occupation was listed as a clerk. The family moved around Maplewood frequently, each time making a small profit on their previous dwelling, sinking the money into a more expensive home.

Dr. W. H. Townsend delivered the baby. Although birth records were not officially required in Missouri until 1910, Dr. Townsend voluntarily registered the birth, listing the child only as "Russell." The family finally decided to name the boy after his father. To avoid confusion at home, they called the boy by his middle name, Ellsworth. Pee Wee told *New Yorker* critic Whitney Balliett,

> I was a late child, and the only one. My mother was forty. She was a very intelligent person. She had been a newspaperwoman in Chicago and she used to read a lot. Being a late child, I was excess baggage. I was like a toy. My parents, who were pretty well off, would say "You want this or that, it's yours." But I never really knew them. Not that they were cold, but they just didn't divulge anything. Someone discovered a few years ago that my father had a lot of brothers. I never knew he had any.

In 1912, Ellsworth was just about to start elementary school when the family moved to Oklahoma. The family probably moved first to Okmulgee, a gas and oil boom town. According to Pee Wee, his father had decided to prospect for gas and Pee Wee said that he had "hit a couple." The family soon moved forty miles east to Muskogee. Oklahoma had boomed following the transition from "Indian Territory," home of the Five Civilized Tribes, to statehood in November, 1907, and Muskogee, the "Queen City," had mushroomed by 1912. Located on the banks of the Arkansas River, which flowed through what had been Cherokee territory, the town nearly doubled in population in one year. Families like the Russells were attracted by the lure of prosperity offered by the burgeoning oil and gas industries.

Mr. Russell worked there first as the manager of another candy store but by 1914, he was employed as the steward of the local Elks' lodge. The club, established in 1900 with a membership of 37, had been very successful. As steward, Charles Russell served as the club's bartender and was also in charge of arranging for the entertainment, a responsiblity that soon would be important to his son.

Six-year-old Ellsworth entered Whittier Elementary school, just a few blocks from his home. He was an outsider, a nervous, shy child. Although of normal height, he was extremely thin, the result of the difficulty he had with digesting solid food. As the school year progressed, he found himself isolated. He became more and more drawn to music. The child dispelled his loneliness by playing piano rolls over and over on the family's pianola. He was given piano lessons, but the repetition of the exercises soon cooled his desire. Ever indulgent, the Russells bought him a drum set complete with bass and snare, xylophone, bells, a triangle and several bird whistles. When that fad faded, they presented him with a violin.

At Whittier Elementary he began to take violin lessons. Pee Wee recalled:

> I was about nine then. Once a month, the proud parents came to hear us. Then there was this big concert. I was so nervous I had the shakes and broke a string. I had to stand there, with everyone waiting, and change the string.
>
> One day, after I'd played in a school recital, I put my violin in the back seat of our car and my mother got in and sat on it. That was the end of my violin career. "Thank God that's over," I said to myself.

But his interest in music was far from over. One night in 1918, his father took him to an Elks event he had arranged in the Sever's Hotel. Alcide "Yellow" Nunez was holding forth with his band, the Louisiana Five. Nunez, one of the first prominent white jazzmen in New Orleans, had been an associate of the legendary trombonist Tom Brown and was a charter member of the Original Dixieland Jazz Band in Chicago. The leader-cornetist, Nick LaRocca fired him for becoming more and more unreliable (Nunez had an alcohol problem) and because the clarinetist insisted on playing lead. After the split, Nunez prospered for a while, making his first records for the short-lived Emerson Record Company under the name of the Louisiana Five, within a few months of the Elks club date and with the same instrumentation that Pee Wee heard.

During the Louisiana Five's 23-month recording career, primarily for Emerson and Columbia, Nunez consistently shirked the traditional New Orleans clarinet role of playing fills and embellishments around a trumpet or cornet lead. Instead, he himself played the lead, with the trombone relegated to background huffing, puffing and tailgate smears. Through the poor acoustic quality of the recordings, Nunez's good, full tone comes through, with a light vibrato and a supple (for the time) rhythmic sense. But the aspect of Nunez's playing that held young Ellsworth enthralled was the thrill of the unexpected: improvisation. Remembering that night more than forty years later, Pee Wee said: "Nunez played the melody and then he got hot and played jazz. That was something. How did he know where he was and where he was going?"

The electrifying experience stayed with Pee Wee all of his life and helped to shape his approach to jazz and clarinet playing. The day after hearing Nunez, the boy begged his parents for a clarinet. One was promptly provided, lessons included. His teacher was Charles Merrill, the reed player in the pit band at the Broadway Theatre and one of the few professional clarinetists in Muskogee. The lessons were conducted at Merrill's house. Although Oklahoma was officially a dry state, Merrill would sneak swigs of homemade corn liquor during the lessons.

> He'd say, "Excuse me, Mr. Ellsworth," He'd go out to the kitchen, pour himself a hooker of corn whiskey—he didn't know at first that I caught on to that—and after two or three lessons

we'd go overtime one, sometimes two, hours. It got so I was ahead of him on lessons."

Living at Charles Merrill's house was a young woman named Miss Lola Merrill, perhaps Merrill's sister or his daughter (although he doesn't appear to have been married). It is quite possible that this is the Lola immortalized in the Mound City Blue Blowers' 1929 recording, "Hello, Lola," one of Pee Wee's most significant early sessions.

Around 1918, the Russells moved into a comfortable house at 1404 Baltimore Avenue. Mrs. Russell soon became friends with the Lyon family, neighbors who lived across the street. Earl Lyon recalled:

> Ella and Aunt Edith would come over and visit with my mother. Mrs. Russell was kind of tall and thin and had some grey in her hair. She was a very gracious kind of person, but very serious-minded. She didn't joke much, and she talked fast. Aunt Edith and she were very close. Aunt Edith—that's what we all called her—was a little shorter than Mrs. Russell and younger. She had a jovial personality. We didn't see much of Mr. Russell. Our family was very religious—still is—and he was a bartender, so his activity was at night. He was a short, round-faced man with squinty eyes, but he always had a smile on his face. He was always very well groomed and dressed.
>
> One time I overheard a conversation between Ella and my mother in which she said that Ellsworth was not getting enough nutrition. "I'm getting kind of worried about him, Ola. All he eats is peanut butter and jam sandwiches," she told my mother. "I just wish he had the appetite that your boys have." I heard mother say, "Well, just have him come over and eat with us anytime." So it wasn't very long before Ellsworth ate a few meals with us. He learned to eat black-eyed peas and buttermilk and cornbread and sow belly, and chicken and noodles on Sunday.

Lyon was three years older than "Ells," and a baseball pitcher, but they became friends. "He was a very quiet boy, very shy and introverted, but he had a lot of initiative and drive. Music was kind of an obsession. Ellsworth and I got along real well because I was interested in learning music too. He was frail but not fragile. My brother and I protected him from the neighborhood bullies." Lyon recalled that Ellsworth was

always dressed up. Mrs. Russell dressed him up like Little Lord Fauntleroy. He had a little collar above his coat and his coat was buttoned up in front. He was always dressed. I never saw him in overalls, I never saw him sweat, I never saw him dirty. She really took care of her boy.

I used to get the other kids to catch for me. One time, I threw the catcher's glove to Ellsworth and asked him to catch for me. But his little hand would shake and I was a little reluctant to throw hard, and I thought I might hit him in the face, so I took the glove and didn't offer it to him anymore. Sports wasn't his forte—music was his niche. He found it early.

He had a bicycle and a street car stopped just a block away and you could ride to town for a token. He would either ride his bicycle or take the streetcar and come home with music, either classical or popular. He was spoiled. He got practically everything he wanted. If he wanted a fiddle, he'd get it; if he wanted drums, he'd get it. He had a Buffet clarinet, a good quality instrument, that he played.

Lyon was also learning to play music. His instrument was the C-melody saxophone and he and Ellsworth would practice together.

He was a natural. He seemed to be advanced. I would go over to Ellsworth's house and my brother would go with me sometimes and pump the player piano. The music room was the first door on the left as you entered their house. It had the player piano up against one wall and a set of drums that extended out in the room. His clarinet was usually out of the case up on top of the piano, ready for use at anytime. He had a record player in the other corner. The room was never disturbed. I think the room was a reminder to him to practice everyday. And he practiced a lot. He was real advanced in his ability to transpose, to improvise, and he could read music real fast. He would play along with the records and practice until he had it down. I wasn't nearly as good with my horn.

The musical stimulation of the new jazz music was everywhere. New hit recordings by the Original Dixieland Jazz Band were eagerly sought by young Russell. Their first, "Livery Stable Blues," released in 1917, sold millions of copies. Larry Shields' clarinet—another early inspiration of Ellsworth's—was prominently featured on most of the group's recordings. The clarinet became the center of the boy's life. "This was for me," he said. "I could play loud. I could play soft . . . Then you'd get on with the fingering and get a little lip, some kind of embouchure, and you felt good when teacher praised you."

Within a little more than a year, Pee Wee had his first opportunity to play in a band. In the summer of 1919, the Perkins Brothers band was scheduled to play the season at Ray Stem's dance pavilion which extended out over the city reservoir next to Agency Hill, a 52-acre park. Fred was the pianist and Eugene the cornetist. Their father managed a Muskogee department store where Mr. Russell had worked as a clerk. It was Ellsworth's first "professional" job, playing from eleven a.m. to midnight for a grand total of $3.00 per night. "It strengthened my lip," he said later.

When classes began in the fall of 1919, Pee Wee attended Central High School, or so his parents believed. He told jazz writer Charles Edward Smith that he would drive his father to work, drop in at school for a class or two, then set out with a friend for a day of "girls, good times and devilment," after which he'd pick up his father and drive home. A major part of the good times for the 13-year-old involved sampling some of the local corn whiskey.

At that time, jazz music—the "devil's music"—was roundly condemned by all "right-thinking" citizens. Pee Wee had equally embraced the clarinet and the romanticized life-style of the jazz musician. While still in high school, he had become

addicted to alcohol, an addiction that would be central to the
development of his personality and his career. Pee Wee's undis-
ciplined behavior caused many family rifts, but it was a music
job that provided the last straw. He was offered a job playing
with the Deep River Jazzband on one of the riverboats which
plied the Arkansas River as it ran through Muskogee. "I needed
long pants for the job," he told Smith. "I'd never worn them in
my life." He borrowed a pair from his father and hid them in the
garage. "I got a ladder up near my window—sneaked out, got the
pants, met the boys and went on this moonlight cruise on an
Arkansas River boat that went out from a little amusement park
about twenty miles from Muskogee." All went well until Pee
Wee spotted his father in the audience. Mr. Russell had followed him all the way onto the boat. He let Pee Wee finish the
job, then took him home. "That started me on the way to Western [Military Academy] . . . it was either that or reform
school," he said.

The Russells may have been more tolerant than others—Pee
Wee's father, after all, had worked as a bartender—but their
son's drinking and hookey playing caused a family upheaval. Pee
Wee told Balliett: "I wanted to study music at the University of
Oklahoma"—he would have been 14 at the time—"but my
aunt—she was living with us—said I was bad and wicked and
persuaded my parents to take me out of high school and send me
to Western Military Academy."

The choice of the academy was an obvious one. It was a well-
established school with an excellent reputation, and it was located in Alton, Illinois, across the Mississippi from St. Louis,
where family friends and relatives still lived. What is surprising
is that the Russells were able to get their son into this prestigious
school at such a young age (most of the other cadets were a year
or two older), and that they could afford the tuition. No matter
the sacrifice, the family decided that a strong dose of military
discipline was just what was needed to bring the boy back into
line.

Still resentful of that decision more than forty years later, Pee Wee told Balliett that the aunt (in all probability Edith Ballard, who had become schoolteacher while in Muskogee) was still alive, adding that his wife "keeps in touch with her, but I won't speak to her."

Ellsworth was transferred from Central High School to Western Military Academy on September 15, 1920. Colonel C. L. Persing, headmaster, noted that Ellsworth was "a very young man during his tenure at Western." His R.O.T.C. record for 1920–21 shows him rated as qualified in bayonet, care of arms and equipment, care of rooms, infantry drill advanced, infantry pack and tent pitching, military courtesy, physical training and tests and wig-wag signaling. But of greatest importance to him was the opportunity to continue to play the clarinet in the school's brass band and orchestra. "They may not have been the best instructors in the world," Pee Wee said, "but they told me what I was doing wrong and set me right. And the discipline was good. I had to learn. It took me a while—it seemed a long time—but finally I made it, up to first clarinet."

It could not have been a very long time because on October 21, 1921, scarcely a year after his registration, Western Military Academy said goodbye to Ellsworth Russell. "The reason for his dismissal," said Colonel Persing, "is lost in the minds of the men who were his teachers at that time—men who were very tolerant of the changing interests of young men . . . as far as we are concerned, on our records he was an exemplary cadet." Years later, the school would declare that Pee Wee and the great Missouri artist Thomas Hart Benton were their two most distinguished non-graduates. Pee Wee summed up the skill he developed during his military career: "I learned one thing: how to get where you're going on time," and, indeed, throughout his career, no matter in how much of a shambles his personal life might be, he was usually the first musician on the bandstand.

His year at Western was, in all probability, the end of Pee Wee's formal education. He claimed to have attended the Uni-

versity of Missouri in Columbia, and this assertion raises the
problem of separating Pee Wee's fantasy from fact. Perhaps
more than most, Pee Wee embellished his life story when he felt
under attack by questioners. These "facts" often got into print
and then became the basis for much of the biographical misin-
formation that has been published about him. The University of
Missouri story, one he repeated frequently, seems to have no
basis in fact. The University has no record of Charles Ellsworth
Russell ever attending any classes there. Charles Edward Smith
wrote, based undoubtedly on information Pee Wee gave him:

> While at Western, he'd been accepted for the university and
> pledged to a fraternity. Orville Knapp played for tea dances at
> fraternity and sorority houses . . . the interlude [at the Uni-
> versity] was brief." [During the 1930s, Knapp was the leader of
> what musicians called a "Mickey Mouse" band.]

On another occasion, Pee Wee recalled that "on Friday afternoons I'd disappear from Columbia [the location of the University, set approximately in the middle of the state], and commute
to St. Louis or Kansas City and spend the weekends playing jazz.
I'd always be back on campus by Thursday." Another time, he
joked: "Sometimes I'd go away on a Saturday and turn up in
college on Tuesday afternoon in time for a Monday class at eight
o'clock."

It is, of course, extremely unlikely that the University of
Missouri would have accepted a 15-year-old drop-out from
Western Military Academy. The "going A.W.O.L" story perhaps may have been more appropriately applied to his dismissal
from the Academy. His wife, Mary, once told the Washington,
D.C., clarinetist Jimmy Hamilton that Pee Wee had not been a
student, but had stayed at a fraternity house because the members liked his playing.

Whatever the truth, when Pee Wee was terminated from
Western his parents moved back to St. Louis, and by 1922 they
were all living at 1244 Goodfellow Street. Pee Wee's father

worked as a clerk at the New Wellington Hotel and sometime later was the steward of the Planters Hotel. The St. Louis to which Pee Wee returned after eight years was a very different city. It was still adjusting to the many changes brought about by the war and the Volstead Act. The automobile had diminished the importance of the Mississippi River to commerce, but St. Louis still seemed far from decline. Despite, or perhaps because of, Prohibition, its many "social clubs" were more lively than ever, jumping to ragtime and jazz rhythms.

Sedalia, Missouri, a little over 200 miles due west of St. Louis, had been the birthplace of ragtime several decades earlier, but St. Louis became its center when Scott Joplin established himself there after achieving his first success with "Maple Leaf Rag" in 1899. St. Louis was one of the many centers throughout the South where the blues and ragtime were elements still in the process of developing into jazz. The city had become a musical melting pot where hobo blues singers from the Mississippi delta and from the Southwestern frontier mingled with ragtime piano professors and the young aggressive New Orleans jazzmen on the riverboats.

Although St. Louis was a segregated city with separate musicians' unions, jazz bands and hot dance bands could be heard throughout the city by 1922. One of the best of the local black bands was lead by Charles Creath. The recordings he made with his Jazz-O-Maniacs for Okeh during that time glimpse the richness of the jazz talent available: trumpeter Leonard Davis; clarinetist Horace Eubanks; trombonist Al Wynn; another local trumpet star, Dewey Jackson, and Creath's brother-in-law, drummer Zutty Singleton.

A local white band causing a sensation in St. Louis at the time was an unlikely "novelty" group, the instrumentation of which was a newspaper-and-comb, a kazoo and a banjo. Taking their name from the city's nickname, the Mound City Blue Blowers consisted of William "Red" McKenzie, more than six years Pee Wee's senior, Dick Slevin and Jack Bland. In February, 1924, the group traveled to Chicago to wax the first of several best selling records for Brunswick. (Their recording of "Arkansas Blues" sold over a million copies.) At their second recording session they included "San," a featured tune of another musician who was making quite a name for himself in St. Louis, Frank Trumbauer. "Whenever Trumbauer played," an early St. Louis musical historian, Dr. Bartlett D. Simms, remembered, "musicians, white and black, were around learning phrases and copying his hot style . . . Trumbauer's solo on 'San' became the standard for all sax players to measure up to." A little more than five years later, Pee Wee would have the opportunity of recording with McKenzie and the Mound City Blue Blowers.

In the early 1920s Pee Wee struck up a friendship with another young musician, trombonist Thomas "Sonny" Lee. Lee had come to St. Louis from Huntsville, Texas, and the boys found they had a lot in common, not the least of which was having a good time and listening to the jazz bands. They frequently sat in with the local musicians, both black and white. One memorable occasion was playing a concert at the Booker T. Washington Theatre with Zutty Singleton. (Later, Lee would record with the Creath band for Okeh.) The jazz activity in St. Louis was not limited to local musicians. The city was headquarters for most of the Mississippi riverboats, including the fabled Streckfus Line. Its flagship, the S.S. *Capitol*, sported a jazz band under the direction of Fate Marable, who also played the ship's calliope. Pee Wee heard the band when it included Louis Armstrong and the great drummer Warren "Baby" Dodds. Henry "Red" Allen was another youthful New Orleans trumpet player who Pee Wee heard on the riverboats. Allen had a particular impact on Pee Wee, and they developed two of the most iconoclastic styles in jazz. Throughout their lives, whenever their careers intersected, they always produced remarkable results together.

Afternoon jobs were plentiful on the Mississippi riverboats and Pee Wee managed to snag a few on the S.S. *St. Paul* and the S.S. *J.S.* Whenever he could, he'd stay to hear the great black bands play at night. "I worked afternoons," Pee Wee explained. "The 'name' bands worked nights. We'd go up to Keokuk and Davenport on the Streckfus Line boats." Word of the young, thin, nervous musician's abilities began to circulate among the local musicians.

By this time, Pee Wee's family had accepted his determination to become a professional musician and the life style that went with it. Charles, Sr., bought him one more instrument, an excellent Conn alto saxophone. Although Pee Wee claimed it cost $375, the top price for the instrument at that time was around $150.

More and more jobs came to the young musician. Sonny and Ellsworth worked together as often as possible. When one of the boys would get a job, he would try to get the other in the band. One they played in from time to time was pianist Herbert Berger's orchestra. Ellsworth was still growing and the older musicians in Berger's band towered over him. Berger started calling him "Pee Wee" because, the clarinetist recalled, "I always seemed to be around a big bunch of bruisers." He shed Ellsworth, and the diminutive nickname stayed with him even after he had grown to nearly six feet tall.

During the summer of 1922 Pee Wee's parents gave him permission to go with the Allen Brothers' tent show, touring Kansas, North and South Dakota, and Iowa. Pee Wee told Charles Edward Smith that the show had an elephant, a steam calliope and a band that included Orville Knapp. At a stop in the tiny town of Moulton, Iowa, just north of the Missouri border, Pee Wee got a wire from Herbert Berger, who was touring in Mexico. Pee Wee recalled,

> He sent me a telegram, asking me to join him. I was a punk kid, but my parents—can you imagine?—said, "Go ahead, good rid-

dance . . ." I wasn't sure I wanted to go at all. I was scared to death, but then I was convinced that they really didn't love me, that they wanted me out of the way. It wasn't natural that they shouldn't try to stop me, or so I thought.

Pee Wee added he "never came back home to live again." The incident, he said, made him "a loner for life."

When I got to Juarez, Berger told me, to my surprise, I wouldn't be working with him but across the street with piano and drums in the Big Kid's Place.

The club boasted the longest bar on the Mexican-American border. "Three days later, there were union troubles and I got fired and joined Berger." Berger's band was playing at the Central Cafe across the street.

Union regulations prevented members of one local from performing in another local's territory. Musicians had to wait for a certain period before being allowed to join the local. The only exception was for a touring band that was "passing through." Apparently, the union's control extended to the Mexican border towns. Juarez was a wild, shoot-'em-up cowboy town. "There'd be shooting in the streets day and night," Pee Wee remembered, "but nobody paid any attention. You'd just duck into a saloon and wait until it was over."

Unfortunately, the youngster engaged in a bit too much saloon shelter.

The day Berger hired me, he gave me a ten dollar advance. That was a lot of money and I went crazy with it. It was a custom in Juarez to hire a kind of cop at night for a dollar and if you got in a scrape he'd clop the other guy with his billy. So I hired one and got drunk and we went to see a bulldog-badger fight, which is the most vicious thing you can imagine. I kept on drinking and finally told the cop to beat it, that I knew the way back to the hotel in El

Paso, across the river. Or I thought I did, because I got lost and had an argument over a tab, and the next thing I was in jail. What a place, Mister! A big room with bars all the way around and bars for a ceiling and a floor like a cesspool and full of the worst cutthroats you ever saw. I was there three days on bread and water before Berger found me and paid ten dollars to get me out.

The young clarinetist soon became something of a local celebrity, at least at the Central Cafe. A photograph of him was permanently enshrined above the bar. It showed a very thin man who looked much older than he was, with black straight hair, parted in the middle and combed back. His prominent nose and ears seemed too large for the long thin face. The deep set dark eyes conveyed a world-weariness and his forehead already was heavily lined. The photograph graced the bar for decades.

After more than ten months with Berger, Pee Wee struck out on his own. He had a job offer in Tucson, but when he arrived he found the union would not let him play without a lengthy waiting period. As usual, he was broke and didn't have enough money to get back to Juarez. He subsisted on a peanuts and water diet and slept in the train depot and on the campus of the state college. Eventually, he managed to work his way back to Juarez, where Berger was glad to rehire him.

The band went on the road almost immediately. They played the state fair in Phoenix, Arizona, then played a night-club job in Tucson. This time there was no trouble with the union, since the band was on tour. A popular vaudeville act, the Six Brown Brothers (none of whom were related), appeared one night at a Tucson theatre. The band consisted of a half-dozen saxophones, playing wildly pyrotechnic arrangements of what was then called "novelty" music. Just before the performance, one of the "brothers" became ill, and Pee Wee was recruited to play the tricky arrangements without a rehearsal. He managed to get through the engagement successfully.

One night, shortly after the Berger band returned to their regular job at the Central Cafe in Juarez, silent-film director James Cruze came in. He was in the area taking location shots for *The Covered Wagon*. Based on a novel by Emerson Hough, the film was released by Paramount in 1923 and was such a hit that it started what became a new film genre: the Western epic. "He liked the band so much," Pee Wee remembered, "he took us back to Hollywood with him. Rudy Wiedoeft was in the band then. We made pictures, I forget what." Obviously, the movies would have been silent. None have been discovered in which the Berger band can be identified. While in Southern California, the band played at the Green Mill in Culver City and frequently provided the music for wild Hollywood parties.

After the visit to the emerging movie capital, then little more than a small desert village, the band returned to St. Louis in the fall of 1922 and took up residency at the swank St. Louis Club, across from St. Louis University. The building was constructed in 1899 with "sumptuous rooms, bowling alley and swimming pool, within a palatial French Renaissance domicile crowned by picturesque gables." An up-and-coming Utah-born trumpet player, Loring "Red" Nichols, came through St. Louis while the band was playing at the club and made the acquaintance of Pee Wee and Sonny Lee.

As the eventful year of 1922 drew to a close, Berger's St. Louis Club Orchestra made a quick trip to New York to record for the Okeh General Phonograph Company. Very few "territory" bands were summoned to the recording capital as early as 1922—in the days before the recording companies started making field trips. That Berger's orchestra was selected indicates the reputation the band had attained by that time.

They recorded six titles in two sessions over a two-day period, then headed back to St. Louis. The acoustic recordings show the band to be well drilled, with a nice "peppy" rhythm. The tunes were all heavily arranged dance music with no jazz solos, although on one of the titles, "The Fuzzy Wuzzy Bird,"

Pee Wee was given the honor of playing two solo notes on clarinet in the coda. Returning as conquering heroes, to their residency at the St. Louis Club, the band played for more than 600 diners at the New Year's Eve ball.

Pee Wee continued to polish his professional skills in dance bands. Whenever the opportunity came to sit in or to substitute for someone, Pee Wee was ready. One of the bands he worked with from time to time was the Ambassador Bell Hops, under the direction of trumpeter Freddie Laufketter and featuring trombonist Vernon Brown. They played at Trimp's ballroom in St. Louis. In 1926, the band recorded for Okeh in St. Louis under the name of Trimp's Ambassador Bell Hops, but by that time Pee Wee had moved on.

In the July 1946 issue of *The Jazz Record*, Dr. Bartlett D. Simms credits Art Gruner, one of the reed men in the Bell-Hops, with influencing Pee Wee. Gruner "showed Pee Wee plenty of stuff on the clarinet," Simms wrote. "Not that Pee Wee wasn't a mighty apt student and always played fine."

When the famous Victor recording artists, the Benson Orchestra of Chicago, came to St. Louis, Pee Wee sat in with them a few times. The band's reed section included Wayne King, later to become famous as the "Waltz King." King took a liking to the young Russell but couldn't understand why he insisted on playing so forcefully. "Why don't you play clarinet," he inquired, "instead of cornet on the clarinet?" The Alcide "Yellow" Nunez influence was still apparent.

By 1924, still a teenager, Pee Wee had already become a successful dance band musician. He had become proficient on the clarinet, bass clarinet, tenor and alto saxophones. He had seen and experienced a great deal more of the world than most boys his age. But, like many of his compatriots, he had become addicted to alcohol. The twenties were roaring by then, fueled by illegal drink. For a musician working in a speakeasy every night, surrounded by good friends and consistently encouraged to join in—to help create—a good time atmosphere, alcohol was a way of life. Although still estranged from his family—there never would be a reconciliation—Pee Wee found himself accepted by the top musicians in St. Louis, men whose opinions and respect meant a great deal to him. Having mastered the fundamental musicianship required for dance band work, he was ready to begin his jazz odyssey.

2 • • • •

A Jazz
Education

*J*azz bands were springing up throughout the Midwest and Southwest, and one of the best was located in Houston, Texas, and led by the reclusive pianist Peck Kelley. The Kelley band was in need of a reed player for an engagement over the summer of 1924 at Sylvan Beach in LaPorte, Texas, on Galveston Bay. Kelley's first choice was the St. Louis sensation Frank Trumbauer, who was then playing and recording regularly with Ray Miller's popular band. Kelley wired Trumbauer, offering the huge sum of $50 a week, but the C-melody saxophonist turned down the offer and—probably at the urging of Sonny Lee—recommended Pee Wee. When Kelley wired him about the job, Pee Wee jumped at the opportunity. Pee Wee recalled:

> Spats and a derby were the vogue, and that's what I was wearing when I got there. Kelley looked at me in the station and didn't say a word. We got in a cab and I could feel him still looking at me, so I rolled down the window and threw the derby out. Kelley laughed and thanked me.

Although hardly a dandy, Pee Wee was always very careful about his appearance in public, no matter how insolvent his estate or derelict his conduct. His black straight hair was neatly groomed and his clothes were always sedate, starched and pressed. Belying his nickname, he had grown to five feet eleven and one-half inches, but he weighed only about 125 pounds. His long thin arms and legs made him seem to be all right angles.

Kelley "took me straight to Goggan's music store and sat down at a piano and started to play," Pee Wee remembered.

> He was marvelous, a kind of stride pianist, and I got panicky. About ten minutes later, a guy named Jack Teagarden walked in, took a trombone off the wall, and *he* started to play. I went over to Peck when they finished and said, "Peck, I'm in over my head. Let me work a week and make my fare home." But I got over it and was with Kelley several months.
>
> Leon Rappolo was in the band and I had at least heard of him before. But why, I wondered, hadn't I heard about these other guys?

The doomed Rappolo—he would shortly be admitted to a mental hospital, where he spent the rest of his life—had been the clarinet star of the New Orleans Rhythm Kings, the band that created a sensation in Chicago at Friar's Inn during the preceding two years. They had been regarded as the best white jazz band in the city, and Rappolo's playing, so much more inventive and rhythmically relaxed than Larry Shields' or Nunez's, had been the main reason why.

It is understandable that Pee Wee was surprised at the sophisticated level of music he heard Kelley play. Few outside Kelley's birthplace, Houston, had ever heard him, since he adamantly refused to record (although a few recordings do exist and were issued after his death in 1981). He was reluctant to travel and rarely left Houston, even when offered lucrative jobs. Kelley's style was startlingly modern for the time, with a harmonic approach far in advance of other jazzmen. Overall his style bore

similarities to the one later developed by Art Tatum. In Kelley's
playing there was a deliberate attempt to break out of the strait-
jacket of harmonic conventions and explore more interesting
chord progressions and harmonic substitutions. All these stylistic trademarks were absorbed by Pee Wee.

Pee Wee also listened to Jack Teagarden and formed a life-
long friendship with the great trombonist. A year younger than
Pee Wee, "Big T" had been playing with Peck's Bad Boys off and
on since he was sixteen. Pee Wee had never seen or heard
anyone play the trombone the way Teagarden did. The big
Texan had invented an entirely new way of playing the instrument. Off the bandstand, Teagarden and Pee Wee became in-
separable. "When I first went from St. Louis to join Peck in
Houston I felt like I was a big shot arriving in a hick town," Pee
Wee told writer and musician Richard Hadlock.

> Texas was like another country, and nobody down there had
> done any recording as we had in St. Louis. But [when Pee Wee
> heard Peck play], I got scared. I had heard good musicians
> around home—Fate Marable, Charlie Creath, Pops Foster,
> Zutty Singleton—but this was a different thing. Peck not only
> played an awful lot of piano, he played so positive and clean. He
> had a "this is mine" style, with plenty of authority. And he
> wasn't like other fast pianists up north, who didn't know the
> blues. Peck played real blues. He and I spent a lot of time that
> summer listening to Bessie Smith records. It was our way of
> going to church.

Pee Wee remembered Peck with affection.

> If you screwed up or were late or something, Peck would look
> sorry instead of mad. He never bawled out anybody. And if you
> had any feelings, you'd wonder how you could do this to such a
> nice guy. Peck wasn't a religious man, but he was like a
> preacher. He'd make you feel conscious of doing wrong. He never
> knew the meaning of the word "ego." He was humble and quiet
> but not meek. Very sincere. You'd wind up working twice as
> hard for him.

The band opened at Sylvan Beach in May or June, 1924. Leon Prima, brother of New Orleans trumpeter Louis Prima, who was to play a big part in Pee Wee's career later, was on trumpet. Jack Teagarden held down the trombone chair.

The summer was idyllic—for a while. After work, the boys would take their instruments down to the beach for jam sessions against the counterpoint of the waves of Galveston Bay. Rappolo would bring his guitar and sing the blues. After a couple of months, Pee Wee abruptly left the band. Another reed player in the band, Don Ellis, remembered that Pee Wee only played in the band a short time after he (Ellis) joined and that the manager fired Pee Wee—or put pressure on Peck to fire him—because he smoked marijuana on the bandstand. If the story is true, it is unusual. Pee Wee's vice was alcohol; he was not known to be a "viper," as musicians called marijuana smokers. The few months were invaluable to Pee Wee's development as a jazz musician. Peck's expanded harmonic concept—the seemingly wrong note made right—helped to open Pee Wee's ears. He also heard and undoubtedly learned from Rappolo during the brief time both were with the band, and he learned a great deal about music from Teagarden. That summer made a life-long impression on him. He was more determined than ever to be a jazz musician.

When he returned to St. Louis, however, opportunities to play jazz were few. He found plenty of work in society bands, among which were Gene Rodemich's band at the Grand Central Theatre and Ray Lodwig's at the Mounds Country Club. Pee Wee's status as a professional musician was indicated in St. Louis directories, which show his address as 1244 Goodfellow Street, the home of his mother. Pee Wee's father died in 1924 at the age of 61, but his mother continued to live at the Goodfellow Street address until 1929, when she was reported working as a maid in the Belcher Hotel. In the 1930 St. Louis City Directory she is listed as working as a dormitory mother at the Girls Home. Pee Wee used his mother's address as a mail drop. Most of the time he was either on the road or sharing an apartment with Sonny Lee. Although living in the same city, Pee Wee continued to isolate himself from his family as much as possible. That may be one reason that travel appealed to him so much during those years.

A trip that proved beneficial to his interest in jazz occurred around the end of 1924 when Elmer Truch, a clarinetist with the Arcadia Serenaders, got a call from New Orleans trumpet player Wingy Manone, who had a job coming up at the Somerset Club, a San Antonio speakeasy and gambling place. Manone had worked with Truch at the Arcadia Ballroom. He asked the clarinetist to join the band and try to find a good tenor saxophone player to bring with him. Truch immediately sought out Pee Wee, and the two men made the trip to Texas together. The band opened on Christmas Day and played through the winter, with Pee Wee on tenor and Truch on clarinet.

Manone, one of the "hottest" white horn men from New Orleans, gained his nickname as a child in New Orleans when a streetcar accident resulted in the loss of his right arm. He was one of the many musicians who traveled extensively throughout the Southwest during this period. While most jazz historians have concentrated on the "up the river" approach, tracing early jazz development from New Orleans to Chicago, they have largely ignored the immense influence on young white jazzmen of itinerant black instrumentalists and blues singers, riding the rails from west Texas through St. Louis to Memphis and beyond. It is no accident that most of the white jazz musicians who developed in the upper Midwest (Art Hodes being a rare exception) lacked the ability to play the blues as convincingly as their brethren who apprenticed to the south and west of St. Louis— Manone, Teagarden and Pee Wee, for example.

In the spring of 1925, Pee Wee returned to St. Louis. As there were no jazz jobs available, he rejoined the Berger band, which had moved from the St. Louis Club (after a fire closed down their venue there) to the city's newest grand hotel, the

Coronado, catering to visiting Presidents, royalty and assorted celebrities. Located near the geographical center of St. Louis, the hotel had 400 rooms and baths in nine floors above its spacious lobby. It cost more than $2,500,000 to build. When the hotel was formally dedicated on December 28, 1923, Ted Jansen's Orchestra, featuring Frank Trumbauer (described in a newspaper account as a trumpeter), performed at the lavish party. The Berger band took over in time for the next New Year's Eve party ($10 per person) in the Green Room, a small lounge off the lobby.

The Victor Talking Machine Company sent a crew and equipment to St. Louis to record local talent, and Berger's band, which had already successfully recorded for Okeh, was high on its list. On November 3, 1924, the band, now called Herbert Berger's Coronado Hotel Orchestra, recorded two tunes, "Gee Flat" and "Shanghai Shuffle." (The latter had been recorded the month before by Fletcher Henderson's band, featuring Louis Armstrong.) Unfortunately neither selection was issued, and nothing from the session appears to have survived in Victor's vaults. On December 10, 1924, the *St. Louis Post-Dispatch* noted that the band (as Herbert's Dance Orchestra) was broadcasting from the hotel on WCK. Another broadcast was reported on January 13, 1925. In the February 1925 issue of *Greater St. Louis* magazine, a publication of the Chamber of Commerce, there appeared an article headlined "Coronado Hotel Orchestra Achieving Fame for Its Melodies":

> The Coronado Hotel Orchestra, which regularly produces Victor records and is featured on radio programs broadcast from the hotel, played at a Members Conference luncheon of the Chamber of Commerce recently . . .
>
> The Coronado Hotel Orchestra is the maker of Victor records and is also featured on the radio by the Stix, Baer and Fuller station, WCK, directly from the Coronado Hotel every Tuesday and Saturday nights, starting at 11 o'clock.
>
> The orchestra is made up of virtually all St. Louisans. It was

organized here several years ago among college students and played exclusively at college functions. Its fame was broadcast and soon the band started on a tour of the country which made it nationally famous.

At Juarez, Mexico, the Coronado orchestra established a record for an engagement at one of the leading hotels. Later it went to Los Angeles, where it added fame to its enviable reputation at the famous Green Hill.

When the Coronado opened its dining and ball rooms for supper dancing, Manager Bradshaw obtained the services of the orchestra. It has met with tremendous success here. Herbert Berger is the director.

Soon, Berger hired the Texas pianist, Terry Shand, so that he could stop playing and devote full time to conducting and managing the band. Damon "Bud" Hassler, who had played in the violin section of the St. Louis Symphony, also joined the band. Hotel owner and architect Preston J. Bradshaw was so pleased with the turn-away business the band had attracted that he decided to build a larger room onto the hotel to accommodate all the patrons.

Pee Wee would not be in the Berger band when it opened the new room. As important as the association with Peck Kelley and Jack Teagarden had been for the young musician, he was soon to fall under an ever greater influence. "Sonny Lee . . . was playing trombone with this band at the Arcadia [Ballroom]," Pee Wee said, "and Sonny used to live at my home. I came home one afternoon, and there was Bix with Sonny in the living room playing Bix records. It gave me a kick—a big thrill to have Bix in my home."

Bix, the legendary "young man with a horn" was already regarded with awe by jazz musicians throughout the Midwest. Born in Davenport, Iowa, Leon Bix Beiderbecke was three years older than Pee Wee. Bix had been the main attraction of the Wolverine Orchestra and had made several recordings for Gennett in Richmond, Indiana, with the Wolverines and under

his own name. His one shot at the "big time," playing with the
Jean Goldkette orchestra from Detroit, had ended abruptly,
and now he was in St. Louis to join a band being put together
by the C-melody sax player, Frank Trumbauer (nicknamed
"Tram").

Although only 24 years old in 1925, Trumbauer had already
earned a reputation as a "musician's musician" in St. Louis. He
joined the Ted Jansen Jazz Band there in 1920, and the solo he
worked out on "San" became a standard for all the saxophone
players in the area, black and white. Now he was taking his own
band into the Arcadia Ballroom, a large hangar-like wooden
structure with a stone façade on Olive Street, three blocks from
the Coronado Hotel.

Pee Wee hadn't met Bix before Sonny Lee introduced them.
"I had heard him in Chicago," Pee Wee said. "There used to be
a band at the Rendez-vous that Charlie Straight had. Those
were speakeasy days, and Bix used to come late and play with
that band. It would sometimes go on to seven or eight in the
morning. But I had never worked with Bix."

Despite their age difference, the young men had a lot in
common. "We hit it right off," Pee Wee said. "We were never
apart for a couple of years—day, night, good, bad, sick, well,
broke, drunk." They came from similar backgrounds and experiences: both had been deeply hurt by their relationship with their
parents, and they felt rejected by their parents' middle-class
world; both had been shunted to schools away from home in an
attempt to straighten them out. They blotted out that world with
a combination of bootleg whiskey and an all-consuming love of
jazz. They found they both had been initially influenced by the
recordings of the Original Dixieland Jazz Band, Bix modeling
himself after the band's cornetist, Nick LaRocca, and Pee Wee
emulating the clarinetist, Larry Shields. They also shared an
interest in contemporary symphonic music, especially compositions by Stravinsky, Debussy and Ravel, finding the unusual
harmonies and progressions exciting and pleasing to the ear at a

time when most musicians and the general public dismissed such music as "ugly." Bix and Pee Wee also were baseball fans, and they had more than a passing interest in girls.

Meanwhile, Sonny Lee left Herbert Berger's band to play at the Arcadia Ballroom in the relief band, which had originated and recorded in New Orleans the year before as the "Original Crescent City Jazzers." When the band traveled north to play in the St. Louis ballroom in the fall of 1924, Wingy Manone had replaced the original cornetist, Sterling Bose. Renamed as the "Arcadian Serenaders," the band alternated sets with the well-established "Arcadia Peacock Orchestra of St. Louis," then entering their third year at the ballroom. By the time Lee joined the Serenaders nearly a year later, Bose had reclaimed the cornet chair. Both the Serenaders and the Peacock Orchestra recorded several sessions for Okeh in 1924 and 1925. That the Trumbauer band did not is a tragic void in jazz recording history.

The Trumbauer band opened at the Arcadia on September 8, 1925, replacing the Peacock Orchestra, for what would be a nine-month engagement. The band played from 8:30 to 12:30 every night but Monday and did a Sunday matinee for younger dancers.

Lee and Pee Wee urged Trumbauer to consider adding Peck Kelley and Jack Teagarden to the band. When Trumbauer expressed interest in the idea, Lee, Pee Wee and Terry Shand turned their attention to convincing Kelley to make the trip to St. Louis. The reclusive pianist agreed, "just this once." According to Pee Wee:

> Peck scared Bix and Trumbauer. They were crazy over him, and we all agreed he had to be in the band. But we couldn't get past the union. We tried everything, even bribing the union man. The money wasn't as important as the music, and we were willing to pay Peck out of our own pockets. Nothing worked. We got a few club jobs for him to meet expenses, but it was a shame Peck wasn't allowed to work that Arcadia job. He was very advanced

harmonically and was just what we wanted. He went home more convinced than ever that it was a mistake to leave.

Teagarden met the same problem with the union and returned to Texas with Kelley.

The job with Trumbauer paid top salaries for the day. Pee Wee received $75 a week, while Bix earned $90 and Trumbauer collected a leader's fee of $125. The personnel changed only slightly during the course of the engagement.

Bix and Pee Wee's friendship and mutual admiration grew. They rented an apartment in Granite City, Illinois, across the river. With all of the jazz activity in St. Louis and in East St. Louis, Bix and Pee Wee took in as much of it as time permitted. George "Pops" Foster, the New Orleans bass player who made St. Louis his base of operations during this time, remembered the two young men:

> On Mondays all the musicians had the day off and we used to all go over there to see who could burn up the most barbecue. They didn't have a regular barbecue, we just dug a hole in the ground, put rocks in, then some wood and got a fire going. We'd cook the barbecue, eat it, and drink a lot of corn whiskey. We never played or jammed together in those days, that all started in New York. We just got together for kicks. The colored and white musicians were just one. We'd stay out all night, drink out of the same bottle, and go out with the same girls. We used to all pile in Bix's car and go over to Kattie Red's in East St. Louis and drink a lot of bad whiskey. It was green whiskey, man, they sure had bad stuff, but none of us ever got sick on it.

After the symphony season began, the boys attended the Friday afternoon matinee concerts at the Odeon Building. Bud Hassler went with them, and through him and the band's former bass player, Anton Casertani, they were introduced to the symphony musicians and to the conductor, Rudolf Ganz. "Bix had a miraculous ear," Pee Wee said.

There'd be certain things he'd hear in modern classical music, like whole tones, and he'd say, why not in a jazz band? Music doesn't have to be put in brackets. Then later it got to be like a fad and everybody did it. We would often order a score of a new classical work, study it, and then request it from the St. Louis Symphony. And we'd get ourselves a box when they did a program we all liked. We'd haunt them to play scores we wanted to hear. Stuff like "The Firebird." We wanted to hear those scores played well.

A clarinetist in the Symphony, Anthony Sarlie, agreed to give Pee Wee lessons. "I used to try to get him to teach me, and I studied with him a little," Pee Wee said. "I wish I had studied more." Indeed, the lessons with Sarlie concluded Pee Wee's formal clarinet studies. In fact, he rarely practiced any of his instruments. His playing was almost always confined to bandstands and jam sessions. He had nevertheless developed enough technique to read and play any of the day's dance band scores and, more important, had all the technique necessary to play whatever came into his head during his improvisations. One never has the feeling that Pee Wee's creative imagination is fettered by a lack of knowledge about his instrument.

Study time with Tony Sarlie suffered on occasion due to Bix and Pee Wee's pursuit of girls and bootleg whiskey. Even with the girls, Bix and Pee Wee hung out together. Pee Wee dated Estelle Shaffner, while Bix dated her younger sister, Ruth. The two cute brunettes loved to dance to the music at the Arcadia. Nothing much developed between Pee Wee and Estelle, but the relationship between Bix and Ruth deepened as 1925 drew to a close.

Bix and Pee Wee found serious drinking competition in several of the other musicians at the Arcadia, Sterling Bose especially. "Bosie, Bix and I made a hangout of a speakeasy run by Joe Hardaway near the Arcadia," Pee Wee wrote in a letter to an English fan, Jeff Atterton:

Bosie was about five foot five and weighed less than a sparrow. Those were the days when gangs were infesting this country, and St. Louis had the toughest ones. As bad, if not worse, than Chicago. These kids used to own planes and pilots, and they would think nothing of going out and bombing each other. That's on the level. I'm not exaggerating at all. Every speak had its silent partners. Hoodlums who simply declared themselves in.

One Saturday night after work, Bix, Bosie and I went to Hardaway's for our usual before bed snorts. We were all pretty stiff, but Bosie was just about out. Quietly out. He was a good boy and no trouble. He sat on a chair with his feet up against an old fashioned pot-bellied stove when three hard characters came in.

They bellied up to the bar, and one of them said out of the side of his mouth, "Give everybody a drink." Joe turned green. He knew this was the mob, and he knew that this was the beginning of their part ownership. So everybody got a drink. Then they repeated it, and everybody got another drink. That is everybody but Bosie. He seemed to be quietly dozing. Then the third time.

With that, Bosie seemed to come to. He pulled himself up, walked over to the three and said to the leader, "Get the hell out, you son of a bitch."

The guy looked down on Bosie, who was no bigger than a minute and said nothing. What the hell was there to say? Bosie repeated himself. Bix and I were weak with fright. Let me add that Bosie accompanied his request with a little finger poking at the guy's middle . . .

The tough guy hesitated, looked around and said to his buddies, "Come on." And they walked out. He must have thought that this little guy who was not afraid of him and obviously not armed must have powerful connections . . . he had to be one of the boys. The very big boys.

The next day, we had a matinee and made it to Hardaway's first for our usual pickup. Bosie walked in full of the shakes and hoping that Joe will let him add a couple to his tab.

First thing Joe says to him was, "Have a drink on me." Bosie was a little startled. This wasn't according to Hoyle. And Joe wasn't known for his generosity. Meanwhile, we said nothing to Bosie. Joe poured the drinks himself. The second, the third, the fourth. Bosie had a foolish look on his face. He knew something was going on, but he couldn't figure it out . . .

Then Joe told him. He couldn't believe it. At first he thought it was a rib. But Bix and I leveled with him, and he knew it was true. The kid had drawn a complete blank about the night before.

Well, his shakes came back worse than ever. A half hour later he was on a train to Chicago and never came back to the Arcadia ballroom.

In January 1926, the band took a weekend off to drive the 150 miles to Carbondale, Illinois, Trumbauer's birthplace. Tram and his band were greeted as heroes, and the engagement at a local Elks' club was an enormous success. The job, however, included a couple brushes with the law. As Pee Wee told the story:

I didn't know what the situation would be there, so I stuck a pint of gin in every pocket. All the other guys did the same. When we got there the guy behind the wheel went flying up the main street ignoring the speed laws.

The police pulled the car over.

When they saw all the booze they thought we were new bootleggers moving in. We weren't. It was just for us. But they hauled us in. We tried to tell them who we were and that we were musicians in town to play a big dance. Nobody would listen to us. But Trumbauer managed to get to the mayor and explained that he was a local kid who had become a big shot and now come home.

The mayor intervened; the musicians were released and even the gin was returned to them. It was then that Bix realized he had forgotten his cornet. A music store owner, dragged away from his dinner, sold Bix a cornet for $20. "Then," Pee Wee recalled, after the dance, we had all our pockets full of more bottles of booze and started to drive home. All of a sudden Bix said he had to return the horn he borrowed. We all told him he had paid for it, but he kept saying, "A drunk I may be, but I'm no thief."

The musicians headed back to the music dealer's house, where their loud entreaties at 3:30 a.m. awoke his wife. "One look at us was enough," Pee Wee said. "She called the police, and when they saw it was us again and with more whiskey, they just locked us up and threw the book at us."

The Arcadia was due to close for the season on May 3, and the famous Victor recording artist, Jean Goldkette, offered Trumbauer and several of his musicians employment over the summer. Goldkette was forming a new unit to play at the Casino—renamed the Blue Lantern for the season—at Hudson Lake, Indiana, while his regular Victor recording band played the summer at Island Lake, Michigan. Impressed with the high quality of the featured band at the Arcadia, he named Trumbauer the musical director of the new unit, to be called "Jean Goldkette's Blue Lantern Orchestra." The band included several mainstays of the regular Goldkette orchestra. To round out the personnel, Trumbauer chose Pee Wee, Bix, Sonny Lee, Dan Gaebe and Dee Orr. It was a tight group. Most of the band roomed in the hotel across from the Casino, but Bix, Pee Wee, Dan Gaebe, Dee Orr and Itzy Riskin all roomed together in a little yellow cottage in back. That cottage has become a jazz legend.

The cottage was hidden behind the hotel, which stood opposite the Casino. Pee Wee and Bix soon discovered a couple of elderly spinsters living on a nearby farm who made their own corn mash. When Dan Gaebe's car—their sole means of transportation—broke down, Bix and Pee Wee pooled their resources, came up with $60 and bought a 1916 Buick roadster. Eddie Condon remembered what happened next.

"Wait until the guys see us with this," Bix said, "We won't have to ride with them anymore, they'll be begging to ride with us."

Pee Wee was dubious, but he agreed.

They drove back to the lake and hid the car in a side road. Just before starting time that night, they sneaked off, got into it and drove to the pavilion. As they reached the entrance and caught the attention of the boys on the stand the motor stopped. It refused to start; the boys had to push it to the cottage.

Next day they got it going and decided to visit the old maids. They got to the cabin, bought a jug, and started back. Halfway home the Buick went dead. They had to find a farmer and hire him to tow it to a garage.

The car was delivered next day at the cottage. "Park it in the backyard," Bix said to the man who towed it. It never ran again. It had a fine mirror and the owners used that while shaving.

Trumbauer appointed Gaebe to be Bix and Pee Wee's "guardian." They nicknamed him "One Shot Dan," and threw the corks from the bottles at him. He took the hint and left them alone. "Our activities," Pee Wee said, "were no worse or no better than any other two young men in prohibition days. We had enough freedom to do what we wanted to, and no stern fathers to stop us."

The orchestra made several radio broadcasts from the Blue Lantern, but none are known to be extant. Nevertheless, the programs attracted crowds to the ballroom, including many of the young Chicago-area musicians who drove the eighty miles from the Windy City to the lake for a night and morning of revelry with Bix, Pee Wee and the rest of the gang. Benny Goodman, Jimmy McPartland, Bud Freeman, Dave Tough and other jazzmen came to hear the band as frequently as possible and to jam after the Casino closed. Jimmy McPartland remembered one night when the Hudson Lake gang walked into the ballroom where he was playing on 63rd Street in Chicago with Frank Teschemacher and Bud Freeman. "Afterwards," Jimmy said, "we went en masse to hear Louis Armstrong at the Sunset

Cafe. Pee Wee hadn't spent much time in Chicago, if any, and it was a real kick for him to hear Louis.

> When we got to the Sunset, who was in the audience but Maurice Ravel. He loved jazz. When Louis finished at 3 a.m., we all went across the street to the Nest, where Jimmie Noone was playing. Louis, Pee Wee, Bix, all of us, including Ravel, went over. And what a jam session. Everybody played. Pee Wee and Teschemacher played together. And Ravel said he was thrilled to hear this great music. Pee Wee and Tesch really impressed Ravel. So did Bix and Louis, of course.

Bud Freeman remembered another occasion around the same time:

> Tesch and I were playing in Husk O'Hare's Wolverines with McPartland at the White City Ballroom in Chicago. One Saturday night we finished and decided to go to Hudson Lake. We got there about six in the morning. We found Bix and Pee Wee were living together. We banged on the door. No response. They were both knocked out, having drunk themselves to sleep. We got into the hut. There was Bix, completely out. We shook Pee Wee. He stirred. And then suddenly he bolted up, very startled, and started swinging, as though he were being attacked. When he realized it was us, he got out his clarinet and we had a jam session. It was the loudest thing I ever heard in my life . . . and Bix never woke up.

After the session they slept and then went with Bix and Pee Wee to the ballroom for their afternoon performance.

> It was amazing how beautifully they played with their terrible hangovers [Freeman wrote]. I suppose they had their little drinks sitting up on the bandstand, but there was one thing about Bix: he might have had a few drinks just to nurse his hangover but I do not recall ever seeing him play drunk. He played so magnificently all the time. Pee Wee was a very good musician, too, something a lot of people don't understand. He could play those difficult tenor parts in the Goldkette band. He was playing tenor that day and asked me to sit in, which I did. He played the

instrument beautifully but I couldn't get a sound out of it. Probably no one but Pee Wee could. When the horn needed pads he
didn't replace them, he just put rubber bands on the keys to bring
them back into place.

Throughout his career, Pee Wee's lack of concern about the
condition of his instrument was a source of wonder to other reed
players.

Trumbauer, attempting to fashion a career for himself,
couldn't take the boys' behavior. Shortly before the end of the
Hudson Lake engagement, Bix and Pee Wee stopped on the way
to the job for a few drinks. Bix noticed the time and rushed to
the Blue Lantern, not noticing that Pee Wee had passed out.
The next night, Pee Wee appeared on time, but Trumbauer told
him that he was fired. Shaking even more than usual, both from
the effects of the day before and the thought of leaving the band,
Pee Wee looked mournfully at Trumbauer and then, like Mel-
ville's *Bartleby the Scrivener,* took his seat. "I wouldn't leave,"
Pee Wee remembered years later, "so that was all right. I just
wasn't paid."

The engagement ended a couple of weeks later, during the
last week of August. Goldkette took his regular band, which now
featured Bix and Trumbauer, on a tour of the East Coast, beginning with a series of dates in New England and winding up at
the Roseland Ballroom opposite Fletcher Henderson. Perhaps
because of his dereliction, Pee Wee was not invited to join, even
if he was willing to work for free.

The association with Bix had been of monumental importance to Pee Wee. In Bix he had found a soul mate. They shared
similar personalities: reticent but genial loners; quiet, friendly
guys, the kind who sat in the corner with little smiles on their
faces, but who took charge whenever jazz was played. And both
approached the music with improvisation first and foremost.
The new sounds of modern composers—dissonance, whole tone
scales and other innovations of the early twentieth century—
were pleasing to their ears and were reflected in their styles.

While others were struck by Bix's golden tone, Pee Wee was most attracted to the cornetist's unusual choice of notes and by his way of improvising on the chords of a tune, rather than merely embellishing or paraphrasing the melody.

Pee Wee always acknowledged this debt to Bix; throughout his life he always named the cornetist as his greatest influence. "I worshipped the man," he said. "I think he's one of the greatest musicians that ever lived. He had more imagination and more thought than anybody else I can think of." In a radio interview, Pee Wee said, "Bix was like a disease. Everything he played I loved."

3 • • • •

New York in
the 1920s

Pee Wee returned to St. Louis from Hudson Lake in September 1926, while Bix and Tram began to gain a national reputation with the Goldkette band. He spent the summer playing in a band with trombonist Vernon Brown at a resort in Whiting, Indiana, on Lake Michigan. After all the jazz excitement of the last two years, it must have been intolerable for Pee Wee to resume playing with dreary hotel bands. In the spring of 1927, shortly after Pee Wee's twenty-first birthday, Sonny Lee left to try his luck in New York, quickly landing a job in the band at the Waldorf-Astoria.

New York, by 1927, had become the jazz capital. Utah-born trumpet player Loring "Red" Nichols had become a considerable success there. He had begun recording as a sideman more than five years earlier, and his output, with both jazz and dance bands, was staggering. His style, very obviously based on Bix's, lacked only the young master's unique tone—and genius. Nichols' tone was cold in comparison with Bix's, and his phrasing was rhythmically stiff and calculated. There were none of the wonderfully adventurous harmonic excursions; none of the

45

flights of inspired improvisational abandon that defined Bix's style.

But with arrangements by "modernist" Fud Livingston, Nichols was successful in achieving a very popular band sound. His group was called the "Five Pennies," although any number of musicians might be involved at any one time. It rapidly became the number one white jazz band in New York, as contrasted to a dance band like Paul Whiteman's. The early recordings under his name included his musical partner, Miff Mole, the leading white trombonist in New York then; Jimmy Dorsey on reeds; and Vic Berton on tympani and drums. They began recording a series of best-selling records for Brunswick in late 1926. With the popularity the records achieved, Nichols suddenly had more work than he could handle in the recording studios, on radio and in live appearances. He organized several units to fulfill all the contracts.

Word went out from the New York local that Nichols was looking for some men. When Sonny Lee heard the news, he talked to Red about Pee Wee. Nichols remembered Pee Wee from the St. Louis days, and had heard of his role in Trumbauer's great Arcadia Orchestra, already a legend among musicians. Pee Wee received a wire from Nichols, offering him a job. Maybe Nichols wasn't Bix, but Pee Wee was happy to have the opportunity to play in a real jazz band again. He left for New York, first stopping off in Chicago to see some of his old buddies. Pianist Jess Stacy, who was active in Chicago from 1925 on, remembered meeting him for the first time then, while he was working at the Midway Gardens with Muggsy Spanier and Frank Teschemacher. "Pee Wee stopped in to see Tesch," he said. "He was just passing through on his way to join Red Nichols in New York. Pee Wee was pretty famous among us musicians." He also struck up a friendship with drummer Dave Tough, who was playing in the band at the Commercial movie palace in Chicago.

Pee Wee arrived in New York on August 14, 1927, and at

first, he was as intimidated as he had been three years earlier when he joined Peck Kelley's Bad Boys. "I went straight to the old Manger Hotel," he told Whitney Balliett,

> and I found a note in my box: "Come to a speakeasy under the Roseland Ballroom." I went over, and there was Red Nichols and Eddie Lang and Miff Mole and Vic Berton. I got panicky again. They told me there'd be a recording date at Brunswick the next morning at nine, and don't be late.
>
> I got there at 8:15. The place was empty except for a handyman. Mole arrived first. He said: "You looked peaked, kid," and opened his trombone case and took out a quart. Everybody had quarts.

The first tune cut was Hoagy Carmichael's first published composition, "Riverboat Shuffle," written for Bix and the Wolverines. The arrangement was by Fud Livingston, whose spiky clarinet style frequently was confused with Pee Wee's at this point. The first solo chorus was by Pee Wee, and we can hear how strong the Beiderbecke influence was, especially if we imagine how those same notes, the same phrasing, would have sounded coming from Bix's cornet. What a contrast to the staid Bix clichés heard in Nichols' solo! Nichols may have had the notes, but Pee Wee had captured the essence. Pee Wee sounds strong and confident, not in the least intimidated. The growls and squawks that became trademarks of his style in later years are not evident on this session or any of the others he made during this New York visit.

We are fortunate to have Pee Wee's first recorded solo available to us in two takes, so that we can hear that both were totally improvised, from first phrase to last. Many musicians on their first "big time" recording date would fall back on solos that had worked in the past, especially if the tune was familiar. Pee Wee must have played "Riverboat Shuffle" innumerable times—with and without Bix—before this recording, yet here he sounds supremely confident in playing whatever fantastic ideas came into his head. The band members congratulated Pee Wee on his performance. When one of them told him his solo was beautiful, he responded, "No, it was just unusual."

Next the band recorded "Eccentric." It had been a clarinet feature for Leon Rappolo with the New Orleans Rhythm Kings at the Friar's Inn in Chicago. Pee Wee starts things off with a searing introduction, followed by a series of exchanges with the brass section that demonstrate an original musical mind bursting with ideas. The last two choruses are ensemble, with Pee Wee playing a perfect clarinet part throughout, weaving in and out, under and over, creating beautiful little counter-melodies, and never getting in anybody's way. He sounds as though he had been playing with the band all his life.

After recording the tune, the band broke for lunch, then came back to resume the day's work with "Ida, Sweet As Apple Cider." Again, in the two takes available we can hear how different was Pee Wee's approach on each one. Pee Wee starts things off again on Fud Livingston's composition, "Feelin' No Pain." Years later, when the recording was played for Pee Wee, he said, "Yes, they were a smart group in some ways, but if it hadn't been for Bix they would probably have never happened. It was all copied from the Wolverines, and it was no fault of Red's if it was good." Pee Wee's performance, especially on "Ida" and "Feelin' No Pain," made quite an impact among jazz musicians. The record established Pee Wee as an important new voice on the New York jazz scene. It was also a big hit with the public, reportedly selling more than a million copies, a huge amount for the time. By March, 1928, Fred Elizalde, pianist, leader of a popular jazz band in London and critic for the English music publication *The Melody Maker,* had heard about the session. His review in the magazine was Pee Wee's first press notice:

> Again as wonderful as ever as our "hot" friends Red Nichols and his Five Pennies, who have recorded in the most modern style two old favorites, "Riverboat Shuffle" and "Eccentric."
>
> I think by now everyone knows the general style of this masterpiece combination, so I will not risk boring you by repeating it.

"Red" Nichols is, as usual, the trumpet; Miff Mole is on trombone, Arthur Schutt at the piano, Adrian Rollini plays bass saxophone, and Vic Berton, cymbal, etc. Pee-Wee Russell (clarinet) and Dick McDonough (guitar), who also "appear" are comparatively new comers, taking the places of Jimmy Dorsey and Eddie Lang. There are also two more trumpets which play second and third to Red in certain movements . . . After the verse (on "Riverboat Shuffle"), Pee Wee takes a great chorus on his clarinet. Style, phrases—don't fail to notice these features.

The Melody Maker of December, 1928, included the following:

Red Nichols and His Five Pennies—that very modern style "hot" combination, have excelled even themselves on "Ida." This is an old favorite revitalised by the treatment it has been given . . . "Feelin' No Pain," though of a different style is every bit as good as its partner. There is a very tuneful "hot" single string guitar by Ed Lang and later Miff has an accompaniment nearly as "hot" as the solo by Rollini on his goofus.

Two weeks later, Pee Wee again recorded with the Nichols gang, this time under Miff Mole's name. Another recording of "Feelin' No Pain" was made, again with a completely different solo by Pee Wee. Less than a week later, they recorded for Columbia under the name of "The Charleston Chasers," and one more recording of the Fud Livingston composition was made. Pee Wee explained it was a common practice in those days for the same band to record the same tune for several different companies. "We'd make some things for one company, then go down the street to another studio and make the same things. We'd just change the name of the band and maybe mix up the order of the solos," he said.

Pee Wee's old pal, Wingy Manone, was in the studio during the Charleston Chasers session and, as he tells it in his colorful autobiography, *Trumpet on the Wing:*

I sat down in the corner to enjoy myself while they knocked themselves out, but the session got to going so good, I got restless.

I had to do something. They were cutting "Feeling No Pain," and I felt no pain, but so good I had to play.

"Do you think I ought to do a little something on this?" I asked Red.

"Sure, how about taking a break right at the first ending?" Red agreed.

I got out my horn, and Red gave us the down beat. Well, not having a chance to get warmed up real good, I spoiled several masters, trying to get my two-bar break just right.

Red and the boys were beginning to get irritated, but they decided to try one more. Man, by that time I was ready steady. When my two bars came up I hit a break that was so solid reet, it knocked those guys off their feet.

Reviewing the records produced at the session, the English magazine, *The Melody Maker*, wrote:

"Sugar Foot Strut" and another Livingston composition, the slightly futuristic "Imagination," composed chiefly of "hot" solos, in which are introduced really wonderful phrases, is one of those which set the standard in small band "hot" style. One of its most interesting features is the now popular use of the saxophone section for working "organ" chords as a background for the "hot" solos. This gives a particularly full effect to the performances without in any way obstructing the soloist. "Imagination," played by the same combination under the name of Miff Mole's Molers, has already been issued on Parlophone. A comparison of the differences in the subtleties of the two recordings is most interesting. Recording excellent both sides.

Wingy and Pee Wee saw a lot of each other during this period. Charles Peterson, a banjo player who later, as a *Life* magazine photographer, played an important role in Pee Wee's career, recalled a job he played with Wingy and Pee Wee around this time:

Wingy and Pee Wee came running . . . with news of a booking at the Brooklyn Rosemont. They were to take the place of the Indiana Five, who were playing an engagement at the Strand

Theatre. Wingy had a five piece band with himself, Pee Wee, me with a borrowed banjo, Johnny Powell on drums and Alec Kramer on piano. The first night was really sensational. The crowd just went nuts. It must have been the first time they had heard any small band jazz like that. We just knocked the Indiana Five right out of the place.

During the fall of 1927, Pee Wee continued to record in Nichols-led or -produced sessions, the records produced appearing with pseudonyms like "The Red Heads" and "Red and Miff's Stompers." Two others, however, were especially important to Pee Wee.

One session that assumed some importance was with Cass Hagan's band. Trained as a violinist, Hagan was born in Edgewater, New Jersey, in 1904. At the end of June, 1927, he took his band into the roof garden of the Park Central Hotel in New York. In addition to Hayton's arrangements, the band's book included charts by the Goldkette and Whiteman bands' great arranger, Bill Challis. Other arrangements were contributed by Artie Schutt, Ken MacComber, Domenico Savino and Ed Sheasby.

In September, the band moved from the Park Central roof garden to the main dining room, located below street level. Pee Wee was added as the featured jazz soloist and on September 30 cut two titles with the band, "Manhattan Mary" and "Broadway." It was not a steady job, however, and he was a featured attraction only sporadically during the engagement. By January, 1928, the band had ended its residency.

When not with the Hagan band or recording with Red Nichols groups, Pee Wee was "on call." Apparently, he substituted for someone in the California Ramblers around this time, according to a recollection by tenor sax player Arthur Rollini. "During the summer of 1927, I subbed a few times on third alto (with the California Ramblers)," Rollini recalled in his autobiography, *Thirty Years with the Big Bands*. "I was only fifteen years old and so was Irving (Babe) Russin, who played fine tenor sax. Pee Wee Russell was on first alto and sometimes when he wasn't there a fellow by the name of Montgomery played first alto with a big, fat tone." A study of Ed Kirkeby's diaries for the period has revealed that, contrary to some reports, Pee Wee never recorded with the California Ramblers.

Bix was in New York at this time, also without a steady job. He had arrived there with the Goldkette band in late August, around the same time as Pee Wee, whose career had improved since the departure from Hudson Lake nearly a year before. Now, he was playing and recording with Red Nichols, the most popular white jazz group in New York, while Bix was the brightest star in Jean Goldkette's "hot" dance band. But no matter how individually successful they had become, the music business in New York was bad for everyone. On September 18 Goldkette disbanded, and Bix, along with the rest of the band's jazz players, joined Adrian Rollini's orchestra at the Club New Yorker. That job lasted only until October 15. Although at liberty, the band's members were not without hope. Paul Whiteman—leader of the most famous, most popular, and top paying dance band in the country—was sending out signals that he was interested in employing most of them.

In the midst of this, on October 25, 1927, Bix and Trumbauer recorded a session for Okeh: three titles by "Bix and His Gang" and two by "Frankie Trumbauer and His Orchestra." All of the titles recorded that day have become hallowed icons of recorded jazz: "Goose Pimples," "Sorry," "Crying All Day," "A Good Man Is Hard To Find," "Since My Best Gal Turned Me Down"—probably Bix's single greatest day in the recording studio.

Pee Wee was added to Trumbauer's band as a guest soloist for one of the titles cut that day: "Crying All Day." His solo follows one of Bix's very greatest. Few musicians would be able to create anything credible after such an incendiary declamation by the cornetist, but Pee Wee's solo continues at the same high level of intensity. Like Bix's, his solo tells a compelling story all its own and in its own very personal language. It was truly a great reunion—musically as well as personally—for the two friends.

The following day, Pee Wee recorded again with Nichols. Trumbauer was present as part of an apparent reciprocal agreement between the two leaders. Pee Wee's solo on "Sugar" is another excellent example of his style at this time. There are few of the strange tones, bent notes, rasps or growls that were to become so commonplace later. Instead, his tone is crystal clear, almost haughty at times, imbued with the same touch of melancholy that made Bix's sound so captivating.

During November, Red Nichols had a half-hour broadcast over the CBS station WMAQ, as part of the "Robert Burns Panatella Country Club" program. On the broadcast of November 11, the Nichols band played a medley of tunes from John Murray Anderson's *Almanac*. In December, Nichols appeared on his own half-hour program on WMAQ at 9:30 p.m. on Fridays. Featured on various programs throughout the month were Arthur Schutt and Dick McDonough. It is not known if Pee Wee appeared on any of these broadcasts, as apparently none were preserved.

Pee Wee and Sonny Lee shared a dingy one-room apartment on West 71st Street. Soon, a young trombone player, Jerry Colonna, later to become famous as a comedian, also chipped in on the rent. At times, Charlie Peterson, the young banjoist, also roomed there. And, when Bix was in town, the apartment served as one of his headquarters. Late in the year, Pee Wee joined Billy Lustig's Scranton Sirens, which included Wingy Manone, trumpet, and Tommy Dorsey, trombone. The band opened at the Kentucky Club. In November, Dorsey left to join Paul Whiteman's orchestra. A CBS studio bandleader, Jimmy Hilliard, remembered working briefly with the Scranton Sirens in Philadelphia during this period. In addition to Pee Wee, the Sirens then included Sonny Lee (temporarily replacing Dorsey). Manone recommended Jack Teagarden to Lustig, and soon Pee Wee was enjoying introducing the trombonist to the New York crowd. Miff Mole had ruled supreme among New York's trombonists, but when they heard Jack Teagarden play "Diane," everyone agreed that a new era had dawned. The Sirens opened for an eight-week engagement at the Roseland Ballroom with a sensational front line of Wingy, Pee Wee and Big T.

Jazz jobs, however, were not regular enough to pay the rent. Peterson had landed a job in Rudy Vallee's dance band, and, when there was an opening for a clarinetist, he mentioned it to Pee Wee. Pee Wee had also received an offer to join Paul Specht's large society band, which needed a "clean up" reed player. Such a musician was expected to play all the reed instruments as the arrangements required. To make the decision between Vallee and Specht, Pee Wee flipped a coin. It came up Specht. He played alto, tenor and soprano saxophones, with an occasional assignment on the bass clarinet. It was a tough job, playing pedestrian dance band arrangements and precious little jazz.

"We had a contract that called for us to make two records a week. We made them at 8:30 in the morning," Pee Wee remembered. No trace has ever been found of these "records," which probably were radio transcriptions, not commercially released discs.

> We also doubled in vaudeville for sixteen weeks. And that meant all over the city. On top of everything, we did a daily broadcast from the Palais Royal at noon.
>
> In those days, we were through about 1 or 1:30 a.m. and had to be up about 7 the next morning. I don't see how any of us survived. Once, during one of Specht's stage things, we did a thing called *Scenes from the South*. One of the songs was "Carry Me Back to Old Virginny." I'll never forget this if I live to be 800. [For this number, all four reedmen moved down from the bandstand and stood in the spotlight.]
>
> We had been running from place to place. You know, play

your things, put your horn under your coat and run for a cab. My clarinet octave key pad must have become loosened or something because we were doing a section specialty on this song when the pad fell out. It started to roll around slowly, just inside the light from the spotlight. I jumped a tenth and kept looking at the fool thing rolling around in the light. Some of the men on the bandstand saw what was happening, and they either laughed out loud or snickered. Nobody was playing melody. Two of the horns were playing harmony, and I'm not sure what I was playing. Anyway, Specht didn't come in for two days, but when he did, I got it from him and good.

So I thought to myself, as a pretty fresh kid will do, "I'll have my revenge." And I borrowed a couple of clarinets from Charlie McLean, who was working in the pit at the Capitol Theatre.

We were going to make a record, and when I arrived, I set up four clarinets, in addition to my alto and all the rest. When it was time to get started, I kept picking up clarinets and playing a little, then saying, "No, that's not quite right."

I kept this up until Specht caught on and chased me out of the studio. Later on, he sent some of the boys out to a saloon on the corner, and they brought me back.

Pee Wee found playing with Specht was too much like work. He described the bandleader as a tough coal miner from Pennsylvania. "He had no sense of humor," Pee Wee said. "I got so mad, if I'd stayed there I'd have killed him." Pee Wee's career with Specht finally ended after another altercation when Pee Wee, flustered and angry, jumped up and yelled, "Fuck me, you quit!" The story was retold among jazz musicians for years.

Meanwhile, Cass Hagan had lined up a cross-country tour for his band, with Red Nichols and Pee Wee as the featured jazz stars. The tour began October 9, 1928, at the Caravan Ballroom in Washington, D.C. The next night, the band bus rolled into New Castle, Pennsylvania, where they played for a dance at Rainbow Terrace Cascades Park. The tour took the band through Ohio, Indiana, Illinois, Iowa, Kansas City, Utah and then on to California. There they opened on December 2 in

Culver City at "The Plantation," a huge hall designed to look like a southern plantation house, columns and all, and owned by the former film star, Roscoe "Fatty" Arbuckle.

The Fatty Arbuckle scandal in 1921 had convinced the general public that Hollywood was Babylon by the Sea, where "anything goes," even though Arbuckle, after being tried three times for a rape and murder of a model, was completely exonerated. Pee Wee however, was happy to find that Hollywood was indeed everything it was supposed to be. He remembered his stay in the film capital as one grand alcoholic party after another. The engagement at the Plantation ended on January 9, 1929.

The Hagan band began its trek eastward to New York with numerous stops along the way. One was in Kansas City, then a center of jazz activity. The great Bennie Moten band was being rhythmically streamlined by the band's New Jersey-born pianist, William "Count" Basie. The town was alive with blues shouters. It had replaced St. Louis as the magnet for jazz musicians throughout the Southwest. Pee Wee took the opportunity, as was his custom, to hear as many of the local bands as possible. And the local musicians were all fascinated by the modern New York sounds produced by the Hagan aggregation. A young trumpet player who found himself in Kansas City at that time was another Pee Wee, George "Pee Wee" Erwin, and Ervin recalled the impact the Hagan band made on him:

> Another band that played the El Torreon (in Kansas City) in 1929, and then stayed around to work dates in the area, was a band from New York, the Cass Hagan Orchestra, a pretty high-powered outfit and my first real contact with an eastern band. I got to know the members pretty well, including Red Nichols, who was with them on second trumpet (and) . . . a star on tenor sax named Pee Wee Russell. This group impressed me quite a bit and was one of the high spots of my Kansas City stay.

When the Hagan band moved on to Pee Wee's hometown, St. Louis, he decided to stay for a while. On January 25, 1929, he was reinstated in the St. Louis musicians' union. During the winter, he played in the Joe Gill band at the Chase Hotel. The youthful pianist in the band was Gordon Jenkins. While there, Pee Wee ran into an old flame from Muskogee, Lola (possibly the daughter of Pee Wee's clarinet teacher). She was "a pretty little thing," remembered Don Ellis, one of the reedmen in Peck's Bad Boys while Pee Wee was in the band in 1924. Ellis remembered her as being from Tulsa originally. He said she was a "groupie," and was hanging around all the guys in the band, not just Pee Wee.

By this time, Pee Wee was certainly regarded as a star by other Midwest musicians. His recordings with Red Nichols were played everywhere. He was one of their own who had made good in the "Big Apple," and his St. Louis pals figured that it was only a matter of time before he returned to the broadcasting and recording center. During his six months in St. Louis, Pee Wee's style underwent a major evolution. The Beiderbecke influence, although it would always remain strong, was less obvious. Pee Wee's playing had become much more assertive and individualistic. By the time Red Nichols called him to return to New York in the summer of 1929, Pee Wee had developed a new instantly identifiable, jazz voice.

4 • • • •

A Unique
Jazz Style

*R*ed Nichols needed Pee Wee for some more pit band work as well as for recordings. Pee Wee made arrangements to return to New York as quickly as possible. He brought Lola with him. Almost immediately upon his return he resumed a busy recording schedule. On June 6, Pee Wee recorded with a band that used the psuedonym "The Whoopee Makers." Ed Kirkeby, the record producer behind the name, wanted "novelty" tunes played in the most trite manner possible. He hired the best available jazz talent in New York for the sessions. This time the band included one of the original Chicagoans who had replaced Bix in the Wolverines, Jimmy McPartland on cornet, and Pee Wee's Texas friend, Jack Teagarden, on trombone. The first two tunes cut were in the familiar "Whoopee Makers" mold, but the third number was not. "It's So Good" would become a classic, due mainly to Teagarden's vocal and for his first chorus solo on trumpet, his only recording on that instrument. It was another great reunion for Pee Wee.

Within a week, Pee Wee and Teagarden were back in the recording studio again, this time as the "Louisiana Rhythm

Kings," for a session that many critics would point to as a great example of "Chicago jazz." Red Nichols was brought in on cornet. The first tune was a vigorous performance of "That Da Da Strain." It was followed by Teagarden's first recording of "Basin Street Blues," a tune he was to be closely associated with for the rest of his long career. The session concluded with a blues called "Last Cent," probably an apt description of the financial status of the band members, with the exception of Nichols, who was always very careful with his change.

Musicologist Gunther Schuller contends that Pee Wee's solo on "Basin Street Blues" caused musicians "considerable puzzlement" at the time. "And, indeed," says Schuller,

> it was for its time as odd a musical statement to come out of the still developing, relatively young jazz language as had ever been heard. At first hearing one of these Russell solos tended to give the impression of a somewhat inept musician, awkward and shy, stumbling and muttering along in a rather directionless fashion. It turns out, however, upon closer inspection that such peculiarities—the unorthodox tone, the halting continuity, the odd note choices—are manifestations of a unique, wondrously self-contained musical personality, which operated almost entirely on its own artistic laws. What appears at first glance to be a complete ignorance of the laws of orthodox voice leading or melodic construction, is really the expression of a musical vision which goes quite beyond such orthodoxies, supersedes them in highly personal and imaginative ways, and creates its own new alien landscape.

In these first recordings made after a year's recording silence, we hear a very different voice. All traces of traditional clarinet conventions are gone. Pee Wee employs many different sounds: at times, his tone is rough or shrill, precariously sliding on and off pitch; at other times, the sound is soft and warm, whispering or full. His improvisations are punctuated with rasps and growls. He constructs new melodic lines, based on the chords of tunes. They are constructed with unusual choices of notes, including the frequent use of flatted fifths, which were not commonly used in jazz improvisation until the bop innovations of the mid-forties. One can hear the influence—perhaps through the lessons Pee Wee took from Professor Sarlie of the St. Louis Symphony—of early twentieth-century symphonic composers like Stravinsky, Debussy, Ravel and Delius in some of his solos. Never "pretty" or sentimental, his iconoclastic conception is more often than not brought down to earth by a loud snort or hoot, impeccably placed, betraying an irrepressible sense of humor. The formation of Pee Wee's style was almost complete. He had found his unique voice, one that would cause both delight and consternation in listeners through the years. There is probably no other style in jazz more controversial than his. Instantly identifiable to even a casual jazz listener, it has been alternately praised and damned by jazz critics and the general public. Barry Ulanov observed that he was merely "an imitator of Jimmie Noone, a constant part of Eddie Condon's two-beat repertory company," a company "always featuring the wry squeaks and sometimes amusing departures from pitch of Pee Wee Russell's clarinet." Others such as Rudi Blesh dismissed his playing outright as "a kind of sad and childish piping," and "musical nonsense set forth in phlegmy, rasping, 'spit' and 'growl' tones."

The critics were not alone in their denunciation. Even other musicians could be vehement on the subject of Pee Wee Russell—and many are to this day. The great New Orleans clarinetist Barney Bigard said, "I used to buy *Down Beat* magazine all the time but once I read that Pee Wee Russell had won a *Down Beat* poll. That did it. I never read that magazine again. Guys like him and that Frank Teschemacher aren't clarinet players to me." On another occasion, Bigard said, "Man, Pee Wee is terrible—He's terrible because he's nothing. You know what people like? They like to see him squirm and make faces—that's what they enjoy about Pee Wee. The guy is nothing plain and simple. What a stinking tone he's got." But to others, Pee Wee was a weaver of magical dreams, an artist, a storyteller, a poet. Pro-Pee Wee critics often reached for literary analogies when discussing his playing, comparing him to Gertrude Stein and James Joyce. The English author and critic Kingsley Amis proclaimed him the "greatest artist since W. B. Yeats."

In many jazz histories Pee Wee was categorized as a "Chicagoan," although he rarely, if ever, played in the Windy City during his formative years. (His stopover on his way to join Red Nichols in the summer of 1927 seems to be the first time he met Jess Stacy, for example, who had been active in Chicago jazz circles for several years before that.) Many dismissed him as a follower of Frank Teschemacher, the hero of the Chicagoans, although Tesch may have learned as much from Pee Wee as Pee Wee learned from him. Pee Wee believed that to be an effective jazz artist, a musician had to develop his own style.

> That is such a good thing, to have a unique style of one's own. Don't copy anybody. Form your own style, if you want to get anywhere. That's what makes me angry when they say I play Chicago style. I made a lot of records with those Chicago boys but I never played there, and as far as Chicago style, well . . . anyway, I got my own style.
>
> One is bound to hear a phrase that gets into the subconscious, and will eventually come out. It may be months from now, but it is something one has absorbed, perhaps unwittingly and it will come out one day. I don't mean that one just sits down and copies phrases one hears. If you ever start to do that, then you are finished, through. Always remain the master of what you are playing. There are thousands of clarinet players like Benny Goodman, but what good are they? Benny had something to give and so have I, I hope, but not the copyists. And that goes for any instrument. What are we in this business for? Surely we are supposed to be creating something—that's jazz. You play something and if it's good you recognize it immediately, but if it's not you won't. Anyway, that's the way I feel about jazz and have always felt. It must be your own.

Pee Wee was quite forceful about jazz historians who held that
his style was not original but based on Frank Teschemacher's.

> I was playing long before he was. Anyway, he's somebody from
> Chicago. I wasn't there. When Frank Teschemacher was in
> short pants, I'd already been to Juarez, Mexico, and to Los An-
> geles and Hollywood before I ever heard of any of these fellows
> like Condon and Teschemacher. I heard of Goodman because he
> was out there [in Hollywood] with Benny Pollack—when *he* was
> in short pants.

On the few occasions when Pee Wee went to Chicago during his
formative years, he sought out his favorite black clarinetists—
Omer Simeon, Jimmie Noone and Johnny Dodds—not the
young kids who he probably regarded as still learning how to play
their instruments. Bud Freeman, Chicago-born, agreed:

> Tesch was not the precursor of Pee Wee. Quite the opposite. We
> all admired Pee Wee. He was already a successful musician. He
> could read anything. He played third alto with all the top bands
> long before we could get into those things. Tesch regarded Pee
> Wee as a fine artist. We all did. People think he was a wild bird,
> but he was a fine musician—and he played the basic reed family.

In a 1939 article, Dave Dexter, Jr., wrote, "Tesch knew Benny
Goodman and Pee Wee Russell well, and also Milt Mesirow.
But only Pee Wee's playing did he condone . . . Russell played
the way Tesch liked." The similarities in their styles is in the
rhythmic shape of some phrases, of which more later (the staccato attack each employed came from a common source: Jimmie
Noone), and their use of bent notes: playing a note slightly sharp
and "bending" it slightly flat—or vice versa. But, as we have
seen, Pee Wee's basic influences came from the Midwest and
Southwest.

Pee Wee's solos sometimes reminded listeners of a trumpet
style, much more forceful and direct and devoid of the intricate
lace usually associated with the clarinet in jazz. Pee Wee once

said that if he could play a different instrument, it would be the cornet. Certainly part of his love for that instrument was instilled by his association with Bix Beiderbecke, but his interest may have begun at age twelve when he first heard Alcide "Yellow" Nunez play the lead on clarinet. Indeed, at various times, Pee Wee would whip out a cornet mute and play the clarinet through it, thus enhancing the "brass" quality all the more. Although some regarded his use of a mute, infrequently though it was, as merely a gimmick, nevertheless he was able to produce some memorable solos using it.

One of the unique qualities of his style—and the most disturbing element to many of the early critics—was his frequent use of vocalization, producing sounds in his throat while playing, resulting in sounds that resembled a human voice: the "rasping," "spit" and "growl" tones that so disgusted Rudi Blesh. "Dirty," "choked" and "clotted" were just a few of the other adjectives used to describe this aspect of his style. Still other critics referred to his playing as full of "croaking," "squawks" and "squeaks."

Vocalization has long been a part of jazz improvisation. The very earliest black brass players' recordings show them to be proficient "growlers." "Wa wa" mute masters like King Oliver, Bubber Miley, Tricky Sam Nanton, Cootie Williams and many others made their instruments seem to "talk." (Clyde McCoy's "Sugar Blues" hit recording commercialized this approach.) Among reed players, Sidney Bechet used a growl to make his improvisations hotter and, in a later jazz era, Ben Webster's style came to depend on a deep breathiness. But among the important white players of his generation, Pee Wee was the only one to make vocalization a basic part of his style.

"The notes he played would sometimes come out of his throat as much as his reed," said Milt Gabler, the owner of Commodore Records and producer of many of Pee Wee's most important recordings.

He made it squeak because he wanted to. When he played in the
low register it was beautiful when he wanted to play that way. It
was the way he expressed himself. He wasn't like Benny Goodman or others and he didn't want to sound like them. He was like
an avant garde painter. He used the unorthodox because he was
an unorthodox person . . .

   Pee Wee's appeal spread in odd directions. At Ryans, Lester
Young would come in and marvel at Pee Wee. You never knew
what direction a note was going to go with Pee Wee, and how he
was going to get out of it. He would build these impossible
phrases and then work his way out of them. Sometimes he'd slam
into a dead end wall and take a pratfall, but he did it with the
grace of a Buster Keaton.

Pee Wee's ensemble playing also reflected a unique musical
approach. The problem is to "fill the holes" left by the trumpet
and trombone, and not get in the way of the other instruments.
Pee Wee's ensemble work is as impressive as his solos. Gunther
Schuller, in *The Swing Era,* wrote that "as individualistic as
Russell was, he was nevertheless one of the best ensemble
players ever." Dick Hadlock in *Jazz Review* concurred:

> Russell is the ensemble musician par excellence. . . . For-
> saking the undulating lines of more conventional Dixieland clari-
> netists, Russell adds a cutting edge to the top of the ensemble
> sound with a powerful but flexible rasping attack. His unusual
> sensitivity to ensemble harmony is a joy to trumpet players, for it
> permits them to depart from the melody without fear of crashing
> head-on into clarinet notes. Russell touches the traditional third
> above the lead note often enough to construct a "proper" clarinet
> part, but more importantly he stretches the ensemble fabric with
> fourths, fifths (this requires an alert trombonist, for the fifth is
> traditionally his territory), sixths, and ninths, while spinning
> elastic counterlines that are closer to second trumpet parts than
> to the arpeggio-dominated filigrees that one is accustomed to
> hearing in Dixieland and military bands. It is largely his skillful

handling of his very personal ensemble role that gives these old . . . recordings . . . an exhilarating vigor undiminished by time.

Although Pee Wee had relatively little formal musical training, he had enough technique to do what he wanted, and at times that could be very complex indeed. Jimmy McPartland said "he could be a very correct player when he wanted to. I had him on a record date in the late '50s, 'Music Man Goes Dixieland.' There were arrangements and Pee Wee played everything just right. Marvelous tone, perfectly straight." Bud Freeman agreed: "He was well trained. . . . He could read music, had a good sound on the clarinet and a good attack and played very good alto."

"When I first heard him on the radio," said cornetist Ruby Braff,

> I thought there was something wrong with my radio. I said "What is this noise—it's not an instrument." That's the first impression someone who doesn't know will get of him. But when you listen to him play, you know how very modern he is. Technically, he was never limited. He had everything he needed. Technique has nothing to do with music and Pee Wee was interested in music, not technique.

Once, speaking with critic Nat Hentoff about another clarinetist who was known as a virtuoso, Pee Wee said,

> I don't mean to sound egocentric, but if I were to practice five or six hours a day for a few weeks, I could have that degree of technical fluency too. But I don't need it for what I want to say. Some players tend to substitute technical bravado for ideas when they run out of imagination. . . . I like to gamble differently— gamble with the inner music and its possibilities. Harmonically, for example. Bix and I had the same feeling about chords. We'd hear something, and say, "That chord just has to be there, whether it's according to Hoyle or not." You have to hear for yourself, and keep trying new ideas.

Rhythmically, he was a "swinger." It is in his phrasing that Teschemacher most clearly emerges, but who influenced whom is still debatable. They both share a peculiarly sharp—almost angular—attack, hallmarks of Jimmie Noone's style also, and both occasionally use very staccato phrases that other clarinetists would have tongued lightly if at all. But by 1929, Pee Wee had gone much further than Tesch ever dared. He was as advanced as Earl Hines, juggling suspended notes while always keeping the beat supple and propulsive. His phrasing was as free as his melodic improvisations. The way he stretched or compressed the beat, sometimes seeming to turn it completely around and then back again, brings to mind the distortions of reality depicted in surrealistic paintings.

But improvisation ("How did he know where he was and where he was going?" Pee Wee wondered in 1919 when he first heard Alcide "Yellow" Nunez in Muskogee's Elks club) was the bedrock of his jazz conception. Many jazz instrumentalists develop a storehouse of phrases and stylistic clichés that they rely upon. Some polish a solo—a variation on a theme—until it is perfected, then play it the same way for the rest of their careers. Others paraphrase the melody, leaving out some notes, changing others, but basically keeping the melody going. A true jazz improviser—an instantaneous composer-performer who creates a totally new melody on the chords of a tune—is rare, even in jazz.

The preservation of spontaneous improvisation was made possible through the development of the phonograph record. It was routine for record companies to have a band play the same tune several times both as insurance against something happening to the delicate wax master, as well as to have a variety of performances from which to select the best for issue. In many cases, the companies issued more than one "take" of a tune. Thus, we are able to study the improvisational nature of the music from one performance to the next, often separated by only a few minutes in the recording studio. Frequently, alternate takes reveal that what appeared to be a wildly hot, improvised solo was actually carefully worked out beforehand—the various takes do not differ appreciably. And in a few cases, solos that seem to be highly polished turn out to be completely off-the-cuff efforts.

The recordings of the 1920s show only a few musicians whose solos were different conceptually from one take to the next. Bix Beiderbecke and Pee Wee Russell are outstanding examples. This ability spontaneously to compose a new melody, sometimes one far superior to the original, led some early critics to the conclusion that the musicians were too unskilled to be able to play the same thing twice. Pee Wee's squeaks and off-pitch tones were considered to be proof that he did not know the rudiments of his instrument.

"I think Pee Wee, with all his nervous playing and without any facility," Bud Freeman said,

> had more to say in the creative sense than any of the technicians and he became a world-famous figure because people would suffer with him. They'd say, "Oh my God, I hope he gets through this chorus," and this was his charisma, a powerful thing, really . . . In another hundred years, if there is another hundred years, people will talk more about Pee Wee's records than about Benny Goodman's. Although Benny was a great artist, he wasn't what we'd call a creative player—he never created any ideas. But you hear Pee Wee today and you know that voice, that it's Pee Wee.

The great jazz pianist Dick Wellstood wrote of

> the miracle of Pee Wee's playing . . . the crabbed, choked, knotted tangle of squawks with which he could create such woodsy freedom, such an enormously roomy private universe.

And Gunther Schuller points out that even within Pee Wee's own musical universe,

> even in this remote terrain, Russell's musical ideas are almost always unpredictable, either in the large or the small form. And

it is this element of the constantly unpredictable that is the most remarkable measure of his talent and uniqueness. To which one needs to add . . . that he was not just some intriguing, freak, oddball eccentric: he was also one of the most touching and human players jazz has known.

Pee Wee told the *New Yorker*'s Whitney Balliett:

You take each solo like it was the last one you were going to play in your life. What notes to hit, and when to hit them—that's the secret. You can make a particular phrase with just one note. Maybe at the end, maybe at the beginning. It's like a little pattern. What will lead in quietly and not be too emphatic. Sometimes I just use a chord that seems wrong to the next guy but I know it is right for me. I usually think about four bars ahead what I am going to play. Sometimes things go wrong, and I have to scramble. If I can make it to the bridge of the tune, I know everything will be all right. I suppose it's not that obnoxious that the average musician would notice.

When I play the blues, mood, frame of mind, enters into it. One day your choice of notes would be melancholy, a blue trend, a drift of blue notes. The next day your choice of notes would be cheerful. Standard tunes are different. Some of them require a legato treatment, and others have sparks of rhythm you have to bring out.

In lots of cases, your solo depends on who you're following. The guy played a great chorus, you say to yourself. How am I going to follow that? I applaud him inwardly, and it becomes a matter of silent pride. Not jealousy, mind you. A kind of competition. So I make myself a guinea pig—what the hell, I'll try something new. All this goes through your mind in a split second. You start and if it sounds good to you, you keep it up and write a little tune of your own.

Fast tempos are good to display your technique, but that's all. You prove you know the chords, but you don't have the time to insert those new little chords you could at slower tempos. Or if you do, they go unnoticed.

In Pee Wee's new style was a deepened understanding of the blues, which gave his voice a soulful quality. While Bix imparted a certain melancholy at times, Pee Wee plumbed the depth of the blues. Pee Wee told a story, a different story each time, sometimes sad, sometimes humorous, sometimes full of wonder and joy. Those who allowed themselves to fall under his spell "became wholly caught up by the *sound* of Pee Wee Russell—by a singing, strange solitary voice that had never been heard in jazz before," wrote Balliett.

> . . . Lyricism is the result of two rather old fashioned qualities—grace and artlessness [Balliett wrote on another occasion]. These appear when a musician like Russell miraculously and unself-consciously translates such blueslike emotions as melancholy, yearning, and restlessness into a certain bent phrase or huskiness of tone with such surpassing timing and clarity that the listener suddenly becomes a transfixed extension of the musician—a transformation that Russell . . . has been mysteriously accomplishing, with increasing refinement and serenity . . .

In 1989, two decades after Pee Wee's death, Nat Hentoff wrote: "The man's acute musicianship, gently mordant imagination, deeply swinging time and that unforgettable tone all came together in solos that, to this day, are lyrical beyond category and fashion." And Eddie Condon's wife, Phyllis, said, "He must have been stimulated to ecstasy every day to play the way he did."

Despite the controversy about his new style, Pee Wee's sound was immediately appreciated by Nichols and other New York musicians in that circle. Throughout 1929, Nichols kept the clarinetist busy with recording sessions and other projects. During the summer the Five Pennies made a film for the Vitaphone Corporation. It was a typical short subject of the day, used as a filler in movie theatres. The band was presented on a bandstand, just as they might appear any night of the week in a glamorous New York night club. The Pennies played their biggest hit, "Ida," and an extremely hot version of "China Boy." Condon sang the vocal on "Who Cares" and "Nobody's Sweetheart." The sound was recorded on 16-inch discs while the cameras captured the musicians' movements.

In addition to visits to recording and film studios, regular night-club performances and radio appearances, Nichols was getting a band together for John Murray Anderson's revue, *Almanac*. The twenty-two piece band accompanied the performers and was featured during the intermission. The revue opened in Boston at the Colonial Theatre on July 28, 1929, running through August 10th. Then it moved to the Erlanger Theatre in New York. Opening night, August 12, was a success, and the show garnered excellent reviews. A couple of weeks later, Nichols recorded the show's hit song, "I May Be Wrong," featuring a gutsy Pee Wee solo.

With tickets still hard to get, the show suddenly closed on October 12. An informative review by Herbert S. Weil, a very active drummer on the New York scene at the time, appeared in the December 1929 edition of the *Melody Maker,* the leading English music magazine:

> *Almanac,*—the musical comedy hit in the pit orchestra of which were featured Red Nichols, Pee Wee Russell (clarinet), Fud Livingston (tenor sax) and Irving Brodsky (piano)—closed on October 12th, much to the surprise of everyone. That the show was due for a long run was a foregone conclusion, as it played to a packed house at nearly every performance. No reason given, but the general opinion seems to be that internal strife among the owners of the show caused it to end.
>
> Red Nichols at present has his own ten piece orchestra in a new restaurant on Broadway and 48th Street called "Hollywood." Among the boys in the band are Pee Wee Russell on sax and clarinet, whom, I am sure, needs no introduction; Eddie Condon on banjo, who was part owner of "McKenzie and Condon's Chicagoans," a hot outfit that did some fine recording for

Parlophone; Gene Krupa, a great hot drummer; and Joe O'Sullivan, who is one of our greatest hot pianists.

The rest of the orchestra is not up to the usual standard of men employed by Red, and I think by the time this is off the press, he will have made some changes.

There is some talk that Red is going into the revival of musical comedy, "Strike Up the Band," which is scheduled to open soon.

While awaiting further work on Broadway, the band had a steady job at the Hollywood, as the review mentioned. Among the band's chores was providing music for the acts, one of which was three acrobatic Greek dancers: two men hurling a woman through the air climaxed the act. "I heard Pee Wee's clarinet quaver as she sailed through the air," Eddie Condon wrote. "We all had the feeling that if we missed a cue she would either keep going or fall on her face."

As foretold in the *Melody Maker,* Pee Wee's next Broadway job with Red Nichol was in the pit band of George and Ira Gershwin's *Strike Up the Band.* The producers had tried out the show the year before in New Jersey, but George S. Kaufman's anti-war satire was too barbed for the audiences. The play was completely rewritten for the 1929 production. *Strike Up the Band* became one of the Gershwin brothers' greatest successes, the score studded with such gems as "I've Got a Crush on You," "Soon," "The Man I Love," and the title song. Nichols assembled the best musicians available for the job, including Glenn Miller, trombone; Pee Wee and Benny Goodman, alto saxes; but Robert Russell Bennett's stiff arrangements left no space for jazz improvisation. The new production tried out in Boston at the Shubert Theatre on December 24, with George Gershwin himself conducting on opening night. The composer, resplendent in white tie and tails with a huge gardenia, also conducted opening night when the show moved to the Times Square Theatre on January 14, 1930. It ran there for 191 performances, and closed on June 28, 1930, killed by the deepening depression. Years later, Pee Wee remembered the electric excitement on opening night when Gershwin raised his baton to thunderous applause.

Nichols and Pee Wee did not stay for the entire run. Their next show was *9:15 Revue,* a short-lived vehicle for the "Sweetheart of Columbia Records," Ruth Etting. In addition to Pee Wee and Nichols, the pit band included jazz violinist Joe Venuti.

The frantic musical activity with Nichols had muted the effects of the stock market crash on October 29. Pee Wee had more work than he could handle. He considered Nichols a pain in the neck, but he dealt with Nichols as he dealt with the rest of the world: avoided contact as much as possible, was laconic in his responses, and usually appeared diffident. But his career as one of the "Pennies" was ending.

Nichols booked a four-week summer tour of New England ballrooms and colleges. The band included Max Kaminsky, trumpet; Bud Freeman, tenor. Kaminsky remembered a lot of friction between the band members and Nichols. "Nichols loathed us and we returned the compliment. Nichols loved the way Bix played and tried to copy him. But now he was playing with men who knew Bix better than he did." Bud Freeman concurred. "It was an unpleasant experience," he said. "Red was not a warm human being. He was very cold, very authoritarian, always the boss." The ill feelings had become so intense that the band was on the verge of quitting when Nichols fired them and so the tour abruptly ended. That was Pee Wee's last association with Nichols. None of the members of that band appear to have performed for Nichols again.

Although Pee Wee rarely said anything negative about any musician, he did not hesitate, many years later, to express his feelings about Nichols. "Red was nothing but a band musician," he said, "and I don't mean a big band musician, I mean a street band musician. Nobody liked him. He produced such a cold sound—technically he had something, but he made a cold sound, and technicians are a dime a dozen. Of course to me Bix had just about everything—feeling, a glorious sound and a unique way of presenting his music. There was really soul in his playing. And I don't want anything else than that." But the association with Nichols had been of great importance to Pee Wee's career. Nichols gave him a showcase: the opportunity to be heard, as often as possible, in sympathetic jazz surroundings. Through his association with the popular bandleader, Pee Wee had gained an enviable reputation among New York musicians. Within just a few years, Pee Wee had grown from a competent dance band professional in the Midwest and Southwest to a jazz soloist with a unique style, featured with the most popular bands in New York. Pee Wee's special voice intrigued other musicians who heard him. By the beginning of the 1930s, Pee Wee's musical universe was complete.

5 • • • •

Depression
Years

*F*or the nation, the stock market crash of October 29, 1929, signaled that the decade-long party was over. Musicians didn't notice the effects of the crash as much as some other segments of the population: no jazz musicians jumped from windows. Many musicians, like Pee Wee, to whom the stock market was as foreign as Sanskrit, were not directly affected and, for a few months at least, everything seemed to go along as usual.

Leaving the security of Red Nichols' pool of players could not have been easy for Pee Wee. Undoubtedly he made his decision to quit more from peer pressure from the rest of the band than from any initiation on his part, even if he was especially sensitive to being ordered about. When Pee Wee was out of work, he worried about ever working again. And now such worries were founded on reality.

On November 14, 1929, Pee Wee recorded with another "Red," an old St. Louis friend, William "Red" McKenzie. In the early 1920s, McKenzie's popular novelty group, the Mound City Blue Blowers, made recordings for Brunswick that were runaway hits. McKenzie had a warm, compelling voice and also played the comb and tissue paper, through which he produced a tone by humming. He dubbed the resulting sound "blue-blowing." Bing Crosby is said to have acknowledged a stylistic debt to McKenzie's vocals. Indeed, there is a similarity in the relaxed approach that both took to a song. McKenzie, however, was more forceful in his other interest: promoting jazz by getting jobs and recording sessions for his friends.

The date McKenzie organized was at that time one of the very few racially integrated recording sessions, which included the star of Fletcher Henderson's band, Coleman Hawkins. There was an unwritten law against race mixing on the bandstand. It seldom occurred even at private jam sessions. While an admiring white musician might on a rare occasion sit in with a black band, the reverse was simply not allowed to happen by club owners and the union. When recording, however, sympathetic white leaders could use black artists, providing the recording company's executives did not object. This time, the mixture produced a stunning contrast of styles on James P. Johnson's hit tune, "If I Could Be with You," which on the record label was titled "One Hour," since Johnson's melody was never directly stated. The lush, romantic lines of Coleman Hawkins contrasted with the spiky, angular constructions of Pee Wee Russell. On the up-tempo "Hello, Lola" the pair's forceful, declamatory voices meshed perfectly.

Pee Wee and Hawkins, the first great jazzman to play the tenor saxophone, had met on several occasions before the session. They developed a life-long mutual admiration society. Pee Wee told Whitney Balliett that during the 20s and 30s he

> lived uptown at night. We heard Elmer Snowden and Luis Russell and Ellington. Once, I went to a ballroom where Fletcher Henderson was. Coleman Hawkins had a bad cold and I sat in for him one set. My God, those scores! They were written in six flats, eight flats, I don't know how many flats. I never saw anything like it. Buster Bailey was in the section next to me, and

after a couple of numbers I told him, "Man, I came up here to
have a good time, not to work. I've had enough. Where's
Hawkins?"

The lack of a piano allowed more elbow room for Pee Wee
and Hawkins, the two main improvisers, and they made the
most of it. On "Hello, Lola," Pee Wee starts things off with a
piping introduction, then Red McKenzie takes over for two choruses, demonstrating that hot jazz can be played on virtually
anything and sound good. Pee Wee comes in with his own inspired composition, including a few notes with just a slight edge
of vocalization attached to them, followed by a wild and stomping Hawkins chorus. Only then does the momentum drop
slightly while Glenn Miller trots out a series of Jack Teagarden's
patented phrases. The swinging rhythm section keeps everything moving and helps to propel the band into a remarkable
final ensemble.

With the razzle-dazzle out of the way, the two masters con-
centrate on Johnson's ballad. This time, Hawkins starts things
out with a beautifully paraphrased introduction. McKenzie follows on comb, and then Hawkins comes in, very serious this
time, elegant, even rhapsodic. The sound of Pee Wee's first few
notes brings us back to earth. The notes are almost jarring,
drenched in a deep blues feeling. The solo abounds in those
"amusing off pitch" notes that some critics complained about,
and there are many strange choices of notes (how strange they
must have sounded then!). Both men tell their own story in their
music: their solos have a forward momentum, they go somewhere. Both solos also express genuine deep feelings—soul, if
you will—that communicate directly with the listener. Instead
of clashing, the styles of these two giants ultimately complement
each other.

"Hello, Lola" was named in honor of Pee Wee's female
companion, who was probably present at the session. For all of
Pee Wee's shyness, Bud Freeman remembers him as quite a
ladies' man. "He always seemed to have a lot of women. I do not

ever recall seeing Pee Wee without a gal. He always had an affair going. He was a very sensuous man. We all were, but Pee Wee did something about it. Pee Wee always liked to be taken care of, you know. Certain women find that attractive in a man."

Although never officially married, Lola and Pee Wee lived together, off and on, for most of the thirties, a time when such arrangements were not usual. They had what could possibly be termed a "love-hate relationship," with frequent displays of physical, mental and emotional abuse. Lola's identity has never been conclusively established. Based on the descriptions of the people who knew her, she may have been Lola Merrill, daughter of Charles Merrill, Pee Wee's first clarinet teacher in Muskogee. But Lola was a relatively common name in Muskogee during the years that Pee Wee lived there. In addition, several other non-Oklahoman Lolas crossed Pee Wee's path during the years, including a Lola Trowbridge, who sang with the Indianans when they alternated with Trumbauer's Goldkette band at the Blue Lantern Inn. Another was vocalist Lola Bard, who sang with Bobby Hackett's band—presumably at Nick's—in 1938. They, however, seem unlikely, since no one recalled Pee Wee's Lola as being musical in the least. According to Phyllis Condon:

> I just knew her as Lola Russell. She was a waitress, a cute blonde with a very broad Southern accent. It was a very unstable marriage, real tempestuous. She cut up his clothes with razor blades, his tux and his ties. She used to throw things. They had some wild times.

Max Kaminsky remembered:

> That Lola, she was terrible. She took off her shoe one night and let me have it. I ducked just in time. Poor Pee Wee would come to work sometimes all cut up—gashes on his face. One night, when he was working, Lola cut up all his clothes with a pair of scissors. Everything he had. The next day, Pee Wee came in to

us and pulled a great big pair of scissors from his hip pocket. He said, "I'm going to get even tonight."

Ernest Anderson, a Madison Avenue advertising executive who became enamored of the jazz he heard nightly at Nick's in the mid 1930s and started producing jazz concerts, was much more blunt about Lola:

> Don't get the idea that she was some sweet Southern belle. She was little more than a prostitute, and Pee Wee was her john. She was always after him for money and things and if he didn't come across, she'd have some of her gangster clients beat him up. That was the society she ran with: gangsters and mobsters. We were always having to protect Pee Wee from her. And she wasn't pretty either. She had a face like a bulldog.

Of course, Pee Wee was not the best mate either. He frequently drank up his salary, leaving little for Lola to run their apartment. But whatever the connubial difficulties, Pee Wee always needed someone to lean on, someone to deal with the everyday requirements. His life was on the bandstand, playing the clarinet.

Although Pee Wee sought to play only jazz, the opportunities in the early 1930s were few. The night club business was bad and the jazz record industry was on the brink of bankruptcy. But he continued to record as a section man in large dance bands. "I worked in a lot of bands and made God knows how many records in New York," Pee Wee told Balliett, naming the bands of Bert Lown, Ray Levy and the Scranton Sirens. He remembered making "Hit-of-the-Week" records, pressed on paper, with Don Voorhees' orchestra. "We did an experiment," he said. "The saxes faced a wall and played against it. We had to turn around to get the beat from Don. Those records were slipped into newspapers or given away, but I don't think they were very successful." Examination of all of the Hit-of-the-Week records by Voorhees has failed to turn up any distinctive reed work. No recordings during the 1930s by Ray Levy or the Scranton Sirens have been documented. None of the many recordings by Bert Lown during this period reveal any hints of Pee Wee. He may have been remembering radio broadcasts or transcriptions.

Two record dates in September 1930 are known. They were special to Pee Wee because they reunited him with Bix. The sessions, one with Hoagy Carmichael's group and the other with Bix's orchestra, would be the last time they would record together. Bix had hoped to put together a band and take it on a tour of Europe. The session was intended as a rehearsal for the tour, but Bix began to experience nervous collapses, brought on by acute alcoholism, and was in no condition to put together anything. Bix recruited his band at various bars and could not turn down any of the friends he ran into. Thus, he had Jimmy Dorsey, Benny Goodman, Bud Freeman and Pee Wee Russell in the reed section. Pee Wee was given a solo only on the dreary "I Don't Mind Walkin' in the Rain," the least interesting of the titles. While both sessions contain some excellent jazz performances, primarily by Goodman, Bix's session has about it an air of desperation. A far cry from the free-wheeling jazz band of "Crying All Day," this band sounds like a typical under-rehearsed commercial dance band of the period. Even the booting "Deep Down South" reminds one of someone painfully forcing a smile.

On the Hoagy Carmichael session a week later Russell did not solo; all the clarinet work is by Jimmy Dorsey. Again, the approach is commercial, reflecting the demands of the time. Pee Wee's unique sound would not be recorded again for more than a year. By that time, Bix, at age 28, would be dead.

Opportunity seemed to elude Pee Wee even when it presented itself. One of Red Nichols' arrangers and reed men, Fud Livingston, had written a tune that he wanted published. Hungry for immediate cash, for ten dollars Fud offered to let Pee

Wee have a percentage of any profits from the tune, but Pee
Wee was broke also. Finally, Fud got the song published. "I'm
Through with Love" was the biggest hit of 1931.

Jack Teagarden was able to get a couple of jazz record sessions in October and November 1931. Both sessions were
integrated—still a rare occurrence even in the recording
studio—by the Harlem stride piano king, Fats Waller. "Big T"
and Fats were Pee Wee's frequent drinking companions, matching him drink for drink as they made the rounds of Harlem night
spots. Even though the records featured the vocal and instrumental cavorting of Teagarden and Waller, Pee Wee had some
chances to shine. He contributed a pithy break on "That's What
I Like About You"; a guttural solo on "I Got the Ritz from the
One I Love," and a sly one on "China Boy." His solo on "Tiger
Rag," however, lost its way; Pee Wee was virtually drowned out
by the orchestra during the latter part of his solo.

During 1932, when even fewer jazz recording sessions took
place, Pee Wee participated in four outstanding sessions with
the great New Orleans trumpet player Henry "Red" Allen. Publisher Irving Mills, the manager of, among others, Duke
Ellington and Cab Calloway, organized the sessions to promote
one of his new talents, singer Billy Banks. The records were
issued under a bewildering assortment of names but are referred
to collectively as the "Rhythmakers" sessions. An additional session was made under the name of Jack Bland and included the
vocals of Chick Bullock. Nearly all the titles recorded by the
Rhythmakers exist in more than one take. Thus we are able to
enjoy the improvisational differences between the performances. The sessions distill the essence of Pee Wee's playing at
the time on both clarinet and tenor saxophone.

Red Allen had as individual a style as Pee Wee, and they
made a great team. The two had played together in jam sessions
in St. Louis when Allen would appear on one of the riverboats
coming up from New Orleans. Listening to the sessions, one is
also struck by the powerful rhythm section of Joe Sullivan, pi-

ano; Eddie Condon, banjo; the Mound City Blue Blowers' Jack Bland, guitar; the New Orleans-style slap bass of Al Morgan; and the former Charles Creath drummer, Zutty Singleton. The first title cut at the April 18, 1932, session, "Bugle Call Rag," is taken at a breakneck tempo. Apparently singer Banks was late for the session and the musicians decided to go ahead and cut an instrumental while they waited. Pee Wee's playing in the ensembles is superb, and he makes the most of the breaks in the opening chorus and in his all too brief solo. On the second tune, "Oh, Peter!" Pee Wee's contribution is limited to a sober introduction of the theme, followed by a rough-hewn vocal by Allen, filling in for the missing Banks. The vocalist arrived in time to record the next tune, "Margie," which features a smooth tenor solo by Pee Wee and his inspired clarinet playing in the concluding ensemble.

A little more than a month later, on May 23, the band assembled again in the studio. Banks made the session on time and reclaimed the vocal on "Oh Peter," which once again opened with a theme statement by Pee Wee. On the two versions of the song cut at this session, however, Pee Wee also plays a solo chorus and makes the most of it. Again, Pee Wee's ensemble playing is a perfect match for Allen's lead. On "Spider Crawl," Pee Wee contributes nice fills behind the vocals as well as a confident blues chorus. His very hot tenor starts things off on the next tune, "Who's Sorry Now," and a blistering clarinet solo after the vocal. On "Take It Slow and Easy," his solo contribution is limited to a perfectly constructed break, and on the session's concluding tune, "Bald Headed Mama," Pee Wee finally gives out with a hoarse croaking solo, his most unconventional one on the date.

On the next session, July 26, another New Orleans slap bass player, George "Pops" Foster replaced Morgan and Fats Waller replaced Sullivan, and an obscure clarinetist, Jimmy Lord, was added. Pee Wee played only tenor sax on this session. On the first tune, "I'd Do Anything for You," he contributed a very aggressive full chorus solo that any tenor saxophonist would have been proud to play. The ending of his solo on the second take is especially distinctive. Pee Wee gets gritty on the next tune, "Mean Old Bed Bug Blues," especially on the first take, constructing another outstanding blues solo. He is not afforded any solo space on the next two tunes, "Yellow Dog Blues," the venerable W. C. Handy composition, and "Yes, Suh!"

The final Rhythmakers session occurred October 8, 1932, and was issued primarily under Jack Bland's name. Billy Banks was not the vocalist—he was replaced by American Recording Company's "house" singer Chick Bullock. Red Allen, however, took the vocal on the first title, "Who Stole the Lock," and the last, "Someone Stole Gabriel's Horn." The rhythm section remained the same except for Frank Froeba, who replaced Waller. Tommy Dorsey was added on trombone, and Happy Caldwell on tenor, with Pee Wee reverting to clarinet. If there were relatively few examples of Pee Wee's unorthodoxy in the preceding sessions, it was no-holds-barred on both takes of his wildly inventive solo on "Who Stole the Lock." The band next cut three takes of "A Shine on Your Shoes," each in a different tempo, with the first, the issued take, the fastest (almost frantic), and the third take, loping. Tommy Dorsey takes a straight solo sharply contrasted in the middle four bars by Pee Wee's "choke" tone in a "throw caution to the winds" statement. Pee Wee plays an interesting introduction to "Someone Stole Gabriel's Horn" and creates a couple startling breaks.

Pee Wee's solo work is outstanding on this series of recordings, and the ensemble interplay with Allen's brilliant trumpet playing also blazes with creativity. It would have been easy for this association to degenerate into caricatures of their wildly original approaches. It might have been difficult for such forceful players to complement each other. Yet, together they weave perfectly meshed ensembles, each matching the other's inventiveness and directness, and sounding as though they had played together all their lives.

Opportunities to record hot jazz were rare during the Depression. Much more representative of what Pee Wee played nightly during this time was the music captured on two sessions with the Adrian Rollini orchestra in July and September, 1933. Pee Wee's role was mostly limited to playing obbligatos against very commercial dance band arrangements. Nevertheless, his work on "Ah, But Is It Love?" and "Dream On" shows great sensitivity. The Bixian melancholy is at work in his obbligatos on the first tune. On "I Gotta Get Up and Go to Work," with a vocal by Red McKenzie, Pee Wee's solo is one of his "off the wall" efforts, from the first unhinged riff to the concluding bent notes. His clarinet impertinently pokes through the heavy arrangement at the beginning and end of "If I Had Somebody To Love." The second session was less jazzy, with Pee Wee confined to codas and under-recorded obbligatos. Pee Wee's status as a featured "hot" player in aggregations such as the Rollini orchestra served him well during the Depression.

Pee Wee spent most of 1933 in New England, playing during the summer with the Payson Re orchestra at the Megansett Tea Room in Falmouth, Massachusetts. The Re band included a young violinist, Bobby Hackett. Re, who played piano in the band, went on to a distinguished dance band career, winding up as the leader of the orchestra at the swank Stork Club in New York for many years. Pee Wee and Hackett became pals immediately. "Pee Wee taught me how to play—and how to drink. He taught me a lot," Hackett said. "Pee Wee used to sing to me and I'd write it out . . . arrangements for the band. One of them, our arrangement of 'Muskrat Ramble,' knocked around for years and turned up . . . in a movie Jimmy Dorsey made. Those were still prohibition days and we used to buy a couple of gallons of alcohol regularly to make our own gin." Hackett, already proficient on the guitar and violin, had been fooling around on the cornet since hearing Louis Armstrong's records. Pee Wee encouraged him and introduced him to Bix's records, an influence that would prove pivotal in the development of Hackett's early style on cornet.

When Re returned to Boston that winter, the union would

not let Pee Wee and Hackett play with his band. But one of Pee Wee's gangster contacts got them a job at the Crescent Club, a mob-owned after-hours joint which was "protected" from the union. Hackett, playing cornet, and Pee Wee, playing as much tenor as clarinet, formed a trio with pianist Teddy Roy and worked there for three months during the winter. One of Pee Wee's fans at the Crescent Club was another young trumpet player, Max Kaminsky. "I used to stop in and hear him play the pop tunes of the day," Kaminsky remembered. "He played them so marvelously."

Kaminsky and Pee Wee traveled to New York in October, 1933, to take part in a recording session snared by Eddie Condon. It was a no-holds-barred date which produced some outstanding sides: the first was Bud Freeman's composition called "The Eel," with a solo by Pee Wee that sounds as though he were strangling the clarinet. Next, Pee Wee took the first chorus of the haunting "Tennessee Twilight" with a sober reworking of the theme. "Madame Dynamite," an up-tempo riff number, was cut next, with the opening solo again going to Pee Wee, and a heartfelt blues chorus by the clarinetist opened the magnificent "Home Cooking." All the men on the date, Max Kaminsky, Floyd O'Brien, Pee Wee, Joe Sullivan, Condon and the great drummer Sid Catlett, sound as though they were determined to make the most of the rare opportunity to get their kind of music on record. Inspiration burns through every solo, none more so than Pee Wee's. But the American Record Corporation decided that out-and-out jazz was not commercially viable. Nothing from that session was issued at the time.

The boys came back a few weeks later and remade "The Eel," with Pee Wee sailing off into the wild blue yonder on his solo, toying with the powerful rhythm section. They also made another take of "Home Cooking," with composer Alex Hill replacing Sullivan on piano. This time they were successful in making a record that the company felt might have enough commercial appeal to release.

A few days before the end of the year, Pee Wee participated in another session that was deemed too uncommercial to issue. Red McKenzie organized it for the American Record Corporation and got his friends Pee Wee and Jack Teagarden, along with trombonist Tommy Dorsey, to accompany him. The unusual instrumentation also included a xylophone played by Red Norvo, and a harp, played by Casper Reardon. Nothing was issued from that session, although one title, "Mean to Me," survived on a test pressing and hints at what a remarkable session it was.

The repeal of prohibition in December 1933 was a blow to jazz. Speakeasies had been the principal place of employment for jazz musicians. Now that bars and clubs had become legitimate, the ambience of danger, the thrill of breaking the law with impunity, were gone. Many former bootleggers became reputable business men overnight, but former gangsters ran most of the surviving night clubs. A few clubs hung on in midtown Manhattan, but it was a far cry from the days of the late twenties when jazz music was a societal rallying point for rebellious youth, much the same way rock was a few generations later. By the early thirties, jazz was a flickering fad, kept alive by small groups of musicians who refused to compromise (or didn't have the ability or opportunity to compromise) by playing in society dance bands, and by even smaller groups of dedicated fans, the first record collectors, researchers and discographers.

The music business was in a slump, and Pee Wee's life was a shambles. His constant drinking, his tempestuous relationship with Lola and their hand-to-mouth existence were taking their toll. His nervousness grew worse. His face, always so expressive, was seized by twitches, at times so bad that Bud Freeman believed Pee Wee had "St. Vitus dance." His hands shook worse than before, his speech became slurred and garbled, and he had trouble keeping his balance. He developed the habit of strategically positioning himself on the bandstand so that he could lean against the piano to avoid falling over. He was dangerously thin and frequently ill, due, in part, to his inability to eat solid food.

He subsisted on milk, canned tomatoes, soft eggs and other easily swallowed foods. Coming back to his apartment from work in the early morning, Pee Wee often helped himself to quarts of milk that had been left on other doorsteps. Mentally, he seemed less able than ever to handle everyday things and came to rely on others to help him with even the simplest chores (this at a time when he continued to construct intricate and advanced improvisations with total assurance). People who had regarded him as an eccentric now began to laugh out loud at his behavior.

For a self-proclaimed "loner," Pee Wee always had someone who acted as his "bodyguard," as John Dengler put it: someone who would look out for him, help him get across the street, open a locked door for him, drive him somewhere. Sonny Lee was probably the first. Bassist Bob Casey, pianist Dick Cary, cornetist Ruby Braff, clarinetist Kenny Davern and many others filled that role through the years. According to Davern:

> It had to do with his own personal selfishness. Having buddies all around was very well planned. Anywhere in the world he would have people strategically placed, like so many drinks. I think that was basically the style of his life. He had a hell of a life. So when I say he was manipulative by nature, by instinct—by survival— he was that. He could get whatever it was that he needed.

Pee Wee had been on his own since sixteen, an anti-social alcoholic surrounded by adults but beyond parental supervision, so it is no wonder that he improvised methods of survival. Many of his adolescent problems were obviously unresolved until he was well into his forties. So, he was a perceived to be a character. "He looks like the sort of person," wrote critic Charles Edward Smith, "about whom anecdotes are told, an attitude he inspires whether he wills it or not." The stories were inevitably at Pee Wee's expense, such as the one Smith (a close friend of Pee Wee's, it should be noted) went on to relate:

> One story told about him concerns the Chicago El, on which tokens were three for a quarter. Passing through the gate, Pee Wee paid a quarter each time, pocketing the two tokens change.

Gradually, they accumulated and he talked it over with an acquaintance. He explained how he got the tokens and said, "Now what do I do with them?"

On the bandstand, Pee Wee also appeared to the audience to be an "oddball." Musicians at that time were either unassuming while playing or engaged in well-rehearsed hokum such as Georg Brunis playing trombone with his foot. But Pee Wee contorted his tall thin frame as he played, his body writhing with each unexpected turn in his improvisation. He hunched his shoulders and bent his back, looking like an elongated question mark. His furrowed face expressed all the agony and torment that shot from his clarinet. And, of course, the squeaks, squawks, bent notes and odd-sounding phrases all added to the picture of the freak (or fraud) that some saw. Others, however, were captured by his wild abandon and saw in Pee Wee a possessed poet. Milt Gabler remarked:

> He was blessed with a very special countenance. When he played his expressions were remarkable, like Dizzy when his face blows up like a cobra. But Pee Wee could make the funniest faces when he played. It was just how his body reacted when he played, although he was a humorous guy. He knew the audience liked those faces or he got to know it . . . so I always felt he made those faces because the crowd liked it. He never got any hands for it. It was the music that they loved, but the facial thing made him seem more of a character.

Phyllis Condon said:

> Pee Wee was very visual. He was more like Buster Keaton while Eddie was more like Chaplin. He didn't communicate verbally all that well. He shook a lot. When he would try to say something, he'd stammer, gesture and search futilely for the right word. Then he'd make those funny faces when he couldn't say what he wanted to. The faces looked like a face from Picasso's blue period . . . he always acted kind of nervous.

Part of it was an act, as Gabler suspected. Kenny Davern said,

A lot of stories about Pee Wee are like stories about other people whose charm and fame is due in part of their reputation to being eccentric. Pee Wee knew what he was doing. He was playing a role that people expected of him. Pee Wee had many ways of protecting himself from the world. His mumbling, for example. People think that's the way he actually spoke. He could speak as clearly as anyone, but when he didn't particularly want to be conversational, he'd mumble a lot of unintelligible nonsense, and everybody would be nodding their heads in agreement. He would create these imaginary conversations. People didn't want to feel they didn't understand so they'd go along. It was a way in which he put a curtain around himself. He had a million defense mechanisms.

"If Pee Wee appears to talk in a garbled voice at times," Charles Edward Smith said, "it is certainly not because he lacks a lucid mind but because he prefers to play it rather than promote it." Indeed, one of the remarkable things about Pee Wee's speech is that through the garbled sounds can be heard well-chosen words, wit and humor, reflecting an acute and disciplined mind. However, when discussing himself, the words did not always reflect the truth: another defense mechanism he used to keep the fans from behind his curtain was to invent "facts" about himself and his background.

For careful listeners like bassist Bill Crow, Pee Wee's speech could be rewarding. "Pee Wee's conversational style mirrored the way he played," wrote Crow in the January, 1990, edition of *Gene Lees Jazzletter*.

He would sidle up to a subject, poke at it tentatively, make several disclaimers about the worthlessness of his opinion, inquire if he'd lost my interest, suggest other possible topics of conversation, and then would dart back to his subject and quickly illuminate it with a few pithy remarks mumbled hastily into his coffee cup. . . . It was always worth the wait [Crow concluded]. His comments were fascinating, and he had a delightful way with a phrase.

To strangers Pee Wee often appeared painfully shy. "You couldn't just go up to him and ask a bunch of questions," Washington, D.C., clarinetist Jimmy Hamilton remarked:

> If you did, the man would panic and he'd get away from you as soon as he could. He couldn't cope with that at all. Every time I spoke with him, if I hadn't seen him in a long time, it was almost as though you had to start all over again. He was afraid of people.
>
> One time at Blues Alley, as he was leaving the bandstand, some people invited him to come over to their table and have a drink. He held up his clarinet and said, "I have to put this someplace," and the people said to just put it on their table. "No, no," Pee Wee told them, "I have to burn it instead." And then he moved on and we ended up in the alley. It was freezing, but he stood out there because he didn't want to be bothered by those people.

"He had a lot of trouble communicating with most people," said Ruby Braff,

> so he played the dumb character that they seemed to want—the drunken, grumpy boozer—and he wasn't that at all. He was very smart, very intelligent, far more intelligent than most of the people he had to work with. When he did communicate, it was through a rather nasal, somethat choked bass-baritone with a strong southwestern accent, a little like that of Jack Teagarden. His speech was punctuated with unusual expletives: "Bang!" and "Ugh!" were favorites. Everyone was called "Chum" in much the same way that Louis Armstrong used "Gate." A favorite expression was "You got it!," meaning "you make the decision," or "it is up to you."

Record collector and researcher Ken Crawford recalled that when he interviewed Pee Wee, "the man never looked at me, just stared at the ground the whole time and kept backing away."

Pee Wee had no ambition to further his career, much less the drive needed to be a leader. He never had a personal manager or a publicity agent. He was content to play as a sideman. "He had no interest or inclination to be a leader of any kind," Phyllis Condon said. "He was very unaggressive. The poet and the artist. Pee Wee was like Jean Cocteau. Reality didn't enter into his plan unless it forced its way in . . . He led a happy but disordered life. His health may not have been that good, but there were all those wonderful creative years of music."

The greatest of those years were to begin soon, but in 1934 the small-group hot jazz styles of the twenties were no longer popular. Gigs were hard to find. Pee Wee subsisted by cadging drinks, eating canned tomatoes and playing in a Cleveland society dance band. It was led by pianist-violinist Austin Wylie and played at a Chinese restaurant, the Golden Pheasant. The band was surprisingly good and at various times during its long life served as a training ground for such young musicians as Claude Thornhill, who was in the band at the same time as Pee Wee, and later the up-and-coming Artie Shaw. In 1934, Bix was dead and jazz was a flickering flame.

6 • • • •

__52nd Street

*I*n New York in 1934 some of the studio musicians who
worked for the radio stations in midtown Manhattan but remembered the fun of the jazz age, were beginning to take mat-
ters into their own hands. "They'd rehearse their shows in the
morning," recalled Ernest Anderson, an advertising executive,

> then everybody would be released until showtime, which was
> usually eight o'clock at night. So they would all go to various bars
> where they'd hang out all day. That's what started Fifty-second
> Street. Places like the Famous Door. The musicians would go in
> there, leave their instruments, sit around, read newspapers,
> drink, play cards until they worked that night. Even the fellows
> who didn't drink—there weren't many—went in there. Fifty-
> second Street between fifth and sixth was between CBS and
> NBC and the fellows made the place like a club house.

Not only was the Famous Door a musicians' hangout, it was
owned by musicians. Pianist and arranger Lennie Hayton and
his manager, Jack Colt, put up $1,000 each and got eight other
studio musicians—(Jimmy Dorsey, Glenn Miller, Jack Jenney,
violinist Harry Bluestone, Mannie Klein, Gordon Jenkins, then-
trombonist Jerry Colonna, and bassist Artie Bernstein)—to contribute $100 each to capitalize a new after-hours club that would

cater to musicians and their friends. With the $2,800 the group opened the club. They named it the "Famous Door," because the door—not the front entrance but one prominently displayed on a small platform next to the bar—had been autographed by each of the musician-owners. The club was located at 35 West 52nd Street, across the street from an already successful club, the Onyx. It was a few steps down from the sidewalk, with the club entrance to the left. Inside was a dark narrow hallway leading into the main room with the bandstand on the right facing the bar. Celebrities enjoyed inscribing the door.

Mannie Klein ran into New Orleans trumpet player Louis Prima on 52nd Street and suggested Prima put a band together for "this little fun joint for musicians." Prima had been playing with the New Orleans clarinetist, Sidney Arodin, but Arodin joined forces with another Crescent City trumpet player, Wingy Manone, at the Knickerbocker Hotel. Pee Wee, who had played with Louis's brother Leon when they both worked with Peck Kelley's Bad Boys, was recruited. Prima was paid $60 a week and the sidemen each got $40—good money in 1935. Opening night was March 1, 1935, and it marked the start of the transformation of Fifty-second Street to "Swing Street." Before the end of the decade, both sides of The Street would be crowded with clubs, one adjacent to another. It became the jazz center of the nation.

The Famous Door was packed on opening night when Prima's band started their theme song, "Way Down Yonder in New Orleans." The musician-owners were congratulating themselves on their success, but at the end of the night, they found they had taken in $79.75 in cash and a drawerful of their own tabs. All the parties had been hosted by them! Then fate stepped in: the Onyx Club burned to the ground that night, and the Famous Door quickly picked up their business.

Prima's group began to attract attention where it counted. Walter Winchell, Ed Sullivan and other columnists began to promote the place. The popularity of the group grew to the point of their being offered a radio program, "Swing It," which was broadcast from the CBS Radio studios nearby at 52nd Street and Madison Avenue. "Ours was the first swing combo to get the full coast-to-coast treatment," Prima remembered. (No air checks, however, are known to exist of these programs.) "We played head arrangements. Paul Douglas, who later became a fine actor, was our announcer."

Prima's approach to jazz was similar to that of Wingy Manone, his New Orleans compatriot. Both modeled their trumpet playing and vocal styles on Louis Armstrong's work and both delivered a showmanship filled with enthusiastic gestures and jivey patter, with its roots deep in the New Orleans tradition that saw no difference between entertainment and jazz. The rhythm was swinging, a propulsive unit playing tight little riff arrangements. Prima became an attraction to the society set for more than musical reasons. "When he shouted, 'Let's have a jubilee,'" said Sam Weiss, a host at the club, "a lot of those sex-starved dames would practically have an orgasm. I think they thought he was shouting 'Let's have an orgy,' in that hoarse, horny voice of his." At the 3:30 closing time, Prima would lead the band off the bandstand through the audience, urging them to join in the parade, and finish up in the middle of the street playing his theme song once again. It may have been old stuff in New Orleans, but it caused a sensation in New York.

While Prima's showmanship brought in the customers, it went less well with genuine jazz fans. "Prima persists in playing identical solos night after night, and indulges in certain tricks that become tiresome after a while," wrote critic John Hammond in a contemporary review, concluding, however, "the night is saved by the magnificent clarinet of Pee Wee Russell." Leonard Feather recalled in his autobiography, *The Jazz Years,* one of his very first American jazz experiences at the Door.

> A highlight of his [Prima's] performance was "Rockin' Chair," for which he would play the dual roles of father and son, hopping from one side of the small stage to the other as he changed

character with each line of the lyric. But here, too, there were
compensatory factors: admirable clarinet by the glum-faced Pee
Wee Russell, seated next to the grand piano, who would occasionally break into an inspired chorus.

Musicians flocked to hear the band—not for Prima's antics
but for Pee Wee's originality. One of them had been one of
Nichols' "Pennies," the tympanist-percussionist, Vic Berton.
Berton had a recording session coming up for Vocalion and,
while he already had clarinetist Matty Matlock lined up, he
invited Pee Wee to participate. It would be Russell's first recording session in fifteen months. Another old face was in the band:
the trumpet player Sterling Bose, who, when last seen, had been
grabbing the first train out of St. Louis after the incident with
the gangsters. Pee Wee was featured on three of the tunes, and
Matlock also tried a few choruses. Pee Wee's hot, intensely
personal statements on "In Blinky Winky Chinky Chinatown,"
"Blue" and "Lonesome and Sorry" contrast vividly with the lim-
pid, routine solos by Matlock.

Pee Wee would soon be recording more frequently. Prima
had a contract with Brunswick Records, signed shortly after his
arrival in New York, and his first session was on September 27,
1934. His original "New Orleans Gang" recording group included New Orleans trombonist Georg Brunis and clarinetist
Arodin. Arodin was replaced by New Orleans reedman Eddie
Miller for the December session. Then, on May 17, 1935, recordings were made by the band that then was playing nightly at
the Door. The performing band had no drummer, but Sam
Weiss was added for recording sessions, probably at the insistence of the producer. Pee Wee became good friends with
McAdams, whose career was cut short by an auto accident. It
was a tight, swinging little band. Even the musical excesses of
Louis Prima lent themselves to the devil-may-care, let's have
fun, feeling of the group. On one of the tunes, "Basin Street
Blues," Pee Wee played the clarinet through a small trumpet
mute, producing a metallic hoarseness that, despite the gim-

mickry, succeeded in making an arresting musical statement. Pee Wee's solos on the Prima recordings, and on the Berton session, were among his outstanding work in that period. These important records have not been properly reissued through the years, probably due to critical distaste for Prima's playing and singing. At the time, the records were very popular. With the New Orleans Gang's latest hits being heard on jukeboxes throughout the country, the Door became even more successful. And as the establishment packed in the customers, other night spots began to open along the Street, narrow rooms filled with tiny tables and as many undersized chairs as possible. In each club was a small bandstand, crammed into a corner. Every now and then the thick smoke would seem to clear a bit and the noise of table talk would be silenced as the patrons strained to hear flights of fancy from Art Tatum, Billie Holiday, Frankie New-ton . . . or Pee Wee Russell.

As their success became more evident, Prima and Pee Wee became targets of the mob. "A couple of hoodlums loaded with knives cornered Prima and me and said they wanted protection money every week—fifty bucks from Prima and twenty-five from me," Pee Wee told Whitney Balliett.

> Well, I didn't want any of that. I'd played a couple of private parties for Lucky Luciano, so I called him. He sent Pretty Am-berg over in a big car with a bodyguard as chauffeur. Prima sat in the back with Amberg, and I sat in front with the bodyguard. Nobody said much, just "Hello" and "Goodbye," and for a week they drove Prima and me from our hotels to a mid-day radio broadcast, back to our hotels, picked us up for work at night, and took us home after. We never saw the protection-money boys again.

Pee Wee always bragged about his associations with big-time gangsters, and told the Pretty Amberg story throughout his life. He liked to tell about using his military school experience to repair gangster's revolvers and tommy guns during prohibition days. Although jazz musicians often were favorites of mobsters, Pee Wee's involvement with them seemed to go deeper than most. It is difficult, however to tell how much he embellished the stories. "He used to tell me how he'd drop off guns into Lake Michigan when he was in Chicago during the 20s," said Kenny Davern. "He knew those guys and they owed him favors."

Once, while recovering from an illness in Boston, Pee Wee was playing on a bandstand behind the bar. Every time one of the bartenders would pass Pee Wee's feet, he would untie the shoelace. To the bartender, this was the height of hilarity. To Pee Wee, always fastidious about his appearance, it was a grave insult. Finally, as the bartender approached again, Pee Wee suddenly lifted his foot and brought it down hard on the bartender's hand. After closing, the bartender was waiting for Pee Wee in the parking lot. "I give you a warning, chum," said Pee Wee. "You touch me too hard and it will be your ass." The bartender slugged him. The next night, a well-known Boston "nothing personal" professional arrived at the bar, fixed his eye on the bartender, then on Pee Wee, then back on the bartender. The bartender threw up his hands in terror, pleading with Pee Wee to make him go away. Finally, Pee Wee relented and sent the hood away. There are other episodes that suggest Pee Wee did have mob protection throughout his life.

In August, Prima asked the Famous Door for a raise and was turned down. He had already decided what he would do if his demands weren't met. The band gave a month's notice on August 7 and shortly thereafter set out for California, where Prima had already lined up backing for a West Coast "Famous Door." Before leaving, Pee Wee telephoned Bobby Hackett in Boston and told him that there would be an opening at the Door in New York. Hackett came down one night and, after seeing the audience's reaction to Prima's slick show, decided he wasn't quite ready for the big time. Another Boston trumpet player, Max Kaminsky, was in the band led by Georg Brunis that replaced Prima.

Prima's five-piece group piled into a car and set out for the West Coast. Their route took them close to Hudson Lake. Pee Wee asked if they could stop there for a few minutes; there was something he wanted to see. The band pulled up at the old Casino. Across the street they saw the cottage. They got out of the cars and walked through the weeds. Behind the cottage was the rusting hulk of a 1916 Buick. Pee Wee pointed to it. "I own half of that," he said proudly. The musicians laughed, but to Pee Wee it was all that remained of the times he had with Bix, its co-owner.

Thanks to the success of the band's records and radio broadcasts, Prima's popularity preceded him to Los Angeles. When the group arrived in September, Prima had no trouble in finding a good location on Vine Street. Opening night at "The Door" was standing room only. Shortly afterward, the band was featured in a short film by Paramount, "Star Reporter in Hollywood, number 2," part of a variety series featuring various new as well as established acts. Prima's band re-created their hit recording of "Chinatown, My Chinatown." Soon Prima was approached with another offer to appear in a short subject, this one for the Vitaphone Corporation. This time, the band was presented as part of a variety show which included trained dogs, a female singing group called the "Six Symphonettes," and rooftop dancer Hal Sherman. The Prima band, dressed in circus costumes, was featured in "Chinatown, My Chinatown" again.

Their next film was a more significant short subject titled "Swing It." The film purported to depict the band's trip from New Orleans to fabled Hollywood, where they overcame initial difficulties and opened a club that was a big success. All the musicians in the band were featured not only on the bandstand but in acting roles as well. "Swing It" was an excellent document of what the band must have sounded and looked like (the nightclub scenes were actually filmed at the Famous Door), even down to the embarrassing musical humor—honking their horns at each other in an apparent attempt to get the last toot in—

committed by both Prima and Pee Wee. Despite such lapses of taste, the film documents the reasons for the band's success. (One of the bit parts in the film was played by Lucille Ball, in her first film performance.)

Prima's popularity now extended far beyond the jazz world. On the threshold of stardom, he was nevertheless concerned about Pee Wee's steady consumption of alcohol. "He drank too much," Prima said. "We tried to get him off it. I had Red Colonna, Jerry's brother, room with him when we settled in Hollywood." According to Kenny Davern:

> They would lock him in his room to keep him from getting drunk. Then he'd bribe the maid to give him the key and he'd go out and stock up. When he got back, he'd have the maid lock him in. Prima never could figure out how he got so drunk all alone in that hotel room. . . . He told me once, that he had a quart of whiskey on each side of his bed so that, in case he woke up in the middle of the night, he wouldn't have to remember which side of the bed he'd left the booze on. He nearly missed a session once because of a hangover. Prima didn't like that.

The first Prima recording session in Los Angeles occurred on February 28, 1936, and included one of his biggest hits, "Sing, Sing, Sing," his own composition. Pee Wee cannot be heard on this recording; the tenor sax solo was taken by a new addition to the band, Joe Catalyne.

Big swing bands were becoming the rage in early 1936 following the success of Benny Goodman's band, and Prima was determined to cash in on the fad. Prima speculated that if his band was popular with five men, imagine how successful they could be with twelve! Pee Wee remembered that Woody Herman

> lent me a couple of arrangements. I think they came from Isham Jones, who had a very good dance band with some nice arrangements by people like Deane Kincaide and Joe Bishop. Prima had a nice show band, but I didn't stay long. I don't like big bands. I

have never liked big bands. You are too regimented. Not that I mind being told what to do, but I can't bear to play that same note every night. All that saxophone section work bores me. I like small groups where you complement what the man next to you is playing. It makes for better jazz than just sitting playing the same routine every night.

The big band recorded for Brunswick in Los Angeles on May 17, 1936. Among the tunes, "Cross Patch" stands out because of Pee Wee's inventive solo. It proved to be, many years later, the record that brought together Pee Wee and one of his most devoted admirers—Kenny Davern. "In 1961," Davern said, "Ruby Braff told Mary [Pee Wee's wife] that I had the original Brunswicks that he made with Louis Prima in 1935. She told Pee Wee and he got very excited and wanted to hear them. Mary called me because Pee Wee was too shy. I came right over through a mountain of snow. I remember him hearing 'Cross Patch,' sitting there and going 'Oooh!' He'd leap up and say 'wow.' Then he'd sit and sock his ear. Winky (his dog) was barking at the high notes and Mary was screaming at Winky. That's how I met him."

Meanwhile in Hollywood, the union began harassing Prima's band members to join. Pee Wee became a full member on June 27, 1936. But the band's sojourn on the West Coast was coming to an end. Prima had received an offer from the Blackhawk in Chicago and made arrangements to leave Hollywood. The big band opened there in late September to generally good reviews. There were regular coast-to-coast broadcasts from the restaurant. A WOR broadcast of October 11 has the band playing their new theme song, "Sing, Sing Sing," followed by "You Can't Pull the Wool Over My Eyes," with a Prima vocal; "Out Where the Blues Begins," with a vocal by their chanteuse, Velma Ray; "The Stars Know That I'm in Love," vocal by Prima; "Tin Roof Blues," "My Heart Wants To Dance," "You're Still in My Dreams" and "Chapel in the Moonlight," all with Ray; and finally "Star Dust" and "I Still Want You," with Prima vocals. The program shows how few jazz numbers were played ("Tin Roof Blues") and just how commercial Prima was attempting to become. Unfortunately, he was saddled with third-rate pop ditties that went nowhere.

"Gargantuan is a good word to describe the sound of the flop made by Louis Prima at Chi's Blackhawk as the old year drew to a close," reported *Down Beat* in January, 1937. "A hit with five men at the cat sessions in the Famous Door in New York and a failure with twelve is a piece of history which only bears out the truth of Connie Mack's philosophy: Never change a winning line-up." The following month *Down Beat* reported that following the Blackhawk engagement, several of the musicians left the band. Pee Wee developed pleurisy, which prevented him from playing for more than two months. Prima hurried back to Los Angeles, where he formed another band.

Even with such an ignominious ending, Pee Wee's period with Prima was fondly remembered by both men. "I was fortunate in having Pee Wee Russell with me," Prima said, "the most fabulous musical mind I have known. He never looked at a note. But the second time I played a lick, he'd play along with me in harmony. The guy seemed to read my mind. I've never run into anybody who had that much talent." Pee Wee said, "It was the happiest period of my life."

7 • • • •

Nick's

Pee Wee stayed on in Chicago and, by April, 1937, was well enough to join a local band, Parker's Playboys. While he had been recuperating, the jazz scene had also been resuscitated. *Down Beat* magazine, one of the first devoted to jazz, had commenced publishing in Chicago in 1934, and by 1936, at the editorial urging of Helen Oakley, had focused its coverage primarily on jazz. It spread the word that jazz, the music that went into eclipse after the rise of singers such as Gene Austin, Russ Columbo and Bing Crosby, was ready for a comeback. The success of the Famous Door in New York had led to an explosion of similar small jazz clubs in the old brownstones on 52nd Street between Fifth and Sixth Avenues. Leon and Eddie's, the Three Deuces (named after a musicians' speakeasy in Chicago), Jimmy Ryans and many others made The Street as much a jazz thoroughfare as Basin Street in New Orleans or State Street in Chicago had been in earlier decades. On August 21, 1935, Benny Goodman's band finally made its breakthrough to popular acceptance at the Palomar Ballroom in Los Angeles, ushering in the Swing Era, the only period when it can be said that jazz was America's popular music.

Unarranged, spontaneous jazz once again became fashion-

able with college students, with the upper-class set from Manhattan's East Side and with artists and intellectuals. Many
youthful converts to swing music became interested in the ori-
gins of jazz. Enthusiasts like John Hammond began to promote
the music and to arrange for jazz recording sessions, sometimes
putting up the money to pay for them. In New York, the rising
jazz tide spread from midtown Manhattan down into Greenwich
Village when Nick Rongetti opened a new club at 140 Seventh
Avenue. A Greenwich Village native, Nick first heard jazz in
1917 when the Original Dixieland Jazz Band came to New York.
At fifteen, he entered college, eventually going to Fordham Med-
ical School to study medicine. Before he graduated, his father
sent him to Italy for four months to learn the family business,
importing. After he returned to the United States, Nick married
and entered Fordham Law School, supporting himself and his
wife by playing piano in various night spots. Although he graduated three years later with a law degree, he never practiced.
Eddie Condon explained: "He didn't pass the bar, he bought it."

In 1921, Nick had started his first speakeasy at 15 Christo-
pher Street. Several moves later and after the repeal of prohibition, "Nick's" opened near Tenth Street and Seventh Avenue.
For years, Nick had his eye on an unprepossessing one-story
building on the corner opposite his club, and finally in 1937, he
was able to lease it. A New Orleans dixieland band led by trumpeter Sharkey Bonano, who had been playing at Nick's old club
since October 1936, opened the new club, alternating sets with
a new band put together by Red McKenzie that featured Pee
Wee Russell. The new club had two upright pianos and a grand
on the large bandstand. The decor attempted to convey the impression of a Tudor mead hall or a hunting lodge with mounted
moose heads impassively surveying the crowd. Three large kegs
protruded from the wall behind the bandstand and a large chan-
delier hovered above the tables. The stage was partitioned from
the rest of the room by two sturdy columns and a drapery that
hung down from the ceiling for a couple feet. On each side of the

bandstand was a metal railing. Wary of falling the three feet from the bandstand to the floor, Pee Wee always positioned himself next to the railing to steady himself as the evening wore on.

The room provided close-quarters seating for 300 at two-by-two-foot tables and a row of wooden booths against a wall of photographs of some of the personalities who frequented the bar. The operation grew to require thirty-five employees. Nick had a liquor stock valued at more than $100,000 that was stored directly behind the bandstand. In addition to jazz; the place became famous for its "sizzling steaks," served on metal platters creating clouds of smoke that mingled with the thick cigarette haze. Nick, almost a caricature of a cigar-smoking tough-talking night-club owner, didn't want to compromise when it came to music for his new club. He wanted real jazz. An amateur pianist himself, Nick could not be dissuaded from playing one of the pianos from time to time. Nevertheless, he knew good jazz when he heard it.

Nick had hired McKenzie to organize the best dixieland band that could be found. One of the first the singer called was his fellow St. Louisan, Pee Wee, still in Chicago with Parker's Playboys. To complete the front line, McKenzie recruited Bobby Hackett (his first job in New York) and Georg Brunis, the veteran trombonist who, as a member of the New Orleans Rhythm Kings, had been a jazz star in Chicago as early as 1922. The rhythm section included Dave Bowman, piano; Eddie Condon, Red's frequent partner, on guitar; Clyde Newcomb, bass; and Johnny Blowers, drums. Although McKenzie organized the band, it was a cooperative outfit, with everyone receiving the same pay. The "leader" changed weekly so that the additional leader money was split equally too.

When Pee Wee walked into the rehearsal, some of the other musicians were shocked. "What are you doing here? We thought you were dead," they told him. Word of his illness in Chicago had spread and grown more dire with each telling. Pee Wee's cadaverous appearance had led to many reports of his demise through the years. With his self-destructive life-style and frail physique, many wondered how he managed to continue to survive.

Opening night at the club was packed. The social set rubbed elbows with the Village jazz fans. Other acts were added to the bill, and soon there were twenty-two performers on the payroll, including 14-year-old Mel Powell improvising brilliantly during intermissions at one of the pianos. But it was the McKenzie group, sparked by Pee Wee's unique style and physical contortions, that captured the attention. After a few weeks, Bonano's outfit left. From then on the McKenzie band, "directed" by Hackett, held forth with only intermission pianists spelling the musicians between sets.

As the musicians settled into the job, they soon found that the drinks were cheaper at 'Julius' across the street at 159 West 10th Street. The little bar, comprising rooms built in 1826, 1845 and 1864, was owned by Pete Pesci, later to be Eddie Condon's partner in the guitarist's own club. As soon as a set at Nick's was over, the musicians ran over to Julius' for a few drinks. "That was where everyone hung out," Phyllis Condon recalled. "Eddie and Pee Wee would take half of the customers from Nick's with them. Pee Wee would always stay out too long between sets over at Julius'." Understandably, this did not sit too well with Nick. He regularly fired several musicians or the entire band, only to hire them back after an hour or two. Pee Wee was fired more often than most. Nick's angry confrontations with the musicians—he called them "muzzlers" when mad—became a standing joke with the regular customers, but they were very upsetting to Pee Wee. "Nick fired Pee Wee hundreds of times," said Ernest Anderson, an advertising executive at Young and Rubicam who began frequenting Nick's almost as soon as it opened.

Everybody would try to talk him into rehiring him because everybody was always trying to protect Pee Wee. He was like a flower

that needed to be sheltered. There was always a certain feeling of
desperation that if Pee Wee couldn't get back in Nick's, he
wouldn't be able to work at all. There was always a tragic thing
about it. So we'd meet in Julius' and try to figure out what to do.
We always knew that if we could get Pee Wee back in the joint
playing the clarinet that Nick would give in, because Nick really
did appreciate Pee Wee's talent.

Anderson recalled a night, however, when they weren't successful in restoring Pee Wee to the bandstand.

Lola had taken Pee Wee's clarinet and hocked it. She'd hock his
clarinet whenever she could, and we didn't even know which
hock shop. We called everybody but we couldn't find a clarinet.
Ed Hall's trio was playing the first set, from 8:30 to 9 and then
Eddie's band—or the band that had Eddie in it; he never had the
band there—would come on. So our strategy was that at exactly 9
o'clock Pee Wee was going to go up and grab Ed Hall's clarinet
and start playing. So he got the clarinet—and it turned out to be
an Albert system. Pee Wee couldn't play it. So there was a big
explosion with Nick, but the next thing you know, Pee Wee is
back.

In fact, Pee Wee played there longer than any of the others, for
more than a decade, with only a few sabbaticals.

The band at Nick's drew the attention of the New York
columnists and attracted a large following among Greenwich
Village bohemians. Artist Stuart Davis, writers John Steinbeck
and John O'Hara, *Life* magazine editor Alexander King, as well
as many musicians, entertainers and celebrities, frequented
Nick's. Condon's witty repartee was an added attraction. Soon,
he was carrying on with the customers at their tables more than
he was on the bandstand playing, another point of contention
with Nick at first.

Pee Wee, Lola, and their dog, a terrier named Nina, moved
into an apartment at 205 West 10th Street, about half a block
from the club, close enough for Pee Wee to go home between
sets. At that time, Greenwich Village was still very much like a

separate small college town, where the intelligentsia not only tolerated what would be regarded as "unsocial" behavior elsewhere but actually encouraged it. Pee Wee fit in well with the "live and let live" philosophy, and he made the Village his home for most of the rest of his life.

John Hammond came into Nick's regularly to scout the talent and was especially interested in Pee Wee's playing. He used him on a Teddy Wilson session he produced on December 17, 1937. The band also included tenor star Chu Berry and trumpeter Hot Lips Page. One of Hammond's discoveries, Sally Gooding, was the vocalist. The sides were never officially issued, but appeared several decades later on a private release. Hammond used Pee Wee only infrequently on sessions after that, but always maintained a high regard for the clarinetist.

Another regular at Nick's was Milt Gabler, whose father owned a small radio and electrical store at 144 East 42nd Street, near the Commodore Hotel. By 1926 young Gabler made his influence felt when the store began stocking records—everything from classical music to Gabler's favorite at the time, vaudevillian clarinetist Ted Lewis. As his interest in jazz matured, he began to collect the latest recordings by Louis Armstrong, Fletcher Henderson and Duke Ellington. Gabler changed the name of the store from the Commodore Radio Corporation to the Commodore Music Shop. "The record department got so big," Gabler said, "we got rid of the radios, we got rid of the sporting goods and tricks and novelties."

Musicians heard of the Commodore's excellent stock of jazz records, including many "race" releases, specially numbered series usually sold only in black areas of the city. Gabler became an avid collector of the old discs from the twenties, and felt that they should be made available again. In 1934 he approached the American Record Corporation (successor to Columbia and Okeh, which also owned the Brunswick and Vocalion labels among others). He was able to get them to re-press some of the items in their catalogue, using the original master stampers, with a special Commodore label, to be sold in his store. A minimum order of 300 copies was required, and Gabler figured it would take about two years to sell them all. Soon, using a mailing list he carefully built up, he put the same records out on the United Hot Clubs of America (UHCA) label. To order the records, the customer joined the "club" for two dollars a year and received notices of the new releases. The names of all the musicians performing on a record were listed on the label, a boon to collectors at a time when discographical information was limited to the French work by Charles Delauney, *Hot Discography,* and a few early researchers like George Hoefer who wrote a column called "Hot Box" for *Down Beat.* Gabler, always a fan of Pee Wee's playing, drew heavily from the Billy Banks sessions with Pee Wee and Red Allen for some of his initial releases.

The United Hot Clubs began to hold jam sessions in recording studios on Sundays. "It was free," Gabler said.

> I would let all my customers know by sending postcards. We'd invite all the critics and writers. The idea was to get musicians work, and get people to know 52nd Street . . . and it worked. They'd do the write-ups of the concerts (let's call them concerts, because that's what they became later—I called them jam sessions), and the prime fellows in getting the musicians to play for nothing were Eddie Condon and John Hammond.

Soon the recording studios were too small to contain the crowds. Gabler moved the jam sessions to clubs on 52nd Street, usually empty Sunday afternoons. One of the first, held at the Famous Door, included Bessie Smith. Soon, others began holding Sunday afternoon jam sessions. Ralph Berton began a series of sessions in the Village. Some entrepreneurs began charging admission although the musicians continued to work gratis. Finally, Gabler moved the sessions to Jimmy Ryans on The Street, where they continued until 1946.

The success of the Commodore-UHCA reissues attracted

the attention of the major companies. John Hammond talked Columbia into initiating a series of reissues with the personnel printed on the label. "That's the end of UHCA," Gabler said. "The majors would take back their masters and do it themselves. The only way for the Commodore Music Shop to stay in business was to start my own label." At the time, there were no small, independent jazz companies doing their own recording. The majors, Victor and American Record Corporation, controlled the market. Starting a specialty jazz label was an untried idea. The importance of the Commodore series, both to Pee Wee's place in jazz history and, in a broader sense, in documenting jazz history itself, can not be overestimated. Without Commodore, the type of music played at Nick's may well have remained undocumented. Through Gabler's efforts and excellent taste, some of the greatest jazz performances were preserved.

Gabler's first session took place the day after one of the great events in jazz history: the January 16, 1938, Benny Goodman concert at Carnegie Hall. The musical high point of that event was Jess Stacy's piano solo on "Sing, Sing, Sing." At the concert, Hackett played the role of Bix Beiderbecke during a "history of jazz" segment, during which he flawlessly re-created Bix's immortal solo on "I'm Comin' Virginia." The following day, Stacy and Hackett joined Pee Wee, Brunis, Condon, bassist Artie Shapiro and drummer George Wettling on Gabler's first recording session. "Everybody was talking about the Goodman concert," Stacy said. "We'd been up late the night before. After Carnegie, a lot of the guys went up to the Savoy to hear Chick Webb. The Carnegie concert was over at 10:45 so the night was still young." Gabler described the recording session:

> I would go down to Nick's and we'd pick the tunes there. That was our rehearsal, so to speak. We didn't want to rehearse in the studio because that was too expensive. We did two blues at that session and since it was the day after the Carnegie concert, we

named them for that. Pee Wee picked "Love Is Just Around the Corner" as his number.

That was my first date. I began to think about how they should be balanced around the mike. When I was in the control room and heard it coming through the monitor speaker, it wasn't what I was hearing when I was actually in the room, to my ears at least. So the engineer said go change it. We only had two mikes then, one near the crack in the piano and one for the band, pretty high up. I moved Brunis back because he was a tough two-fisted player who blew loud. Bobby wasn't that forceful on trumpet so I put him behind Pee Wee. I didn't just line them up. That's the secret of recording, to get them the right distance apart so they sound in balance to the mike.

I remember I made a special effort not to lose the rhythm section. A lot of guys put the drums so far back you could hardly hear them. I always liked to balance it so you could hear the guitar just topping the afterbeat of the drum. The drum has no real pitch to it, like chords, so if you balance it so the guitar is just a little higher than the sock cymbal it doesn't give you that chonk chonk sound. You get a nice relaxed ring instead. They blend. That way the bass and all the rest come together just the way you heard a band live. I think we achieved a nice rhythm balance on "Love Is Just Around the Corner" and the others on that session.

The records were released under Condon's name—the first so issued in more than four years—while the two blues, "Carnegie Drag" and "Carnegie Jump," appeared on a 12-inch release (another gamble on Gabler's part, since the larger format rarely had been used for any music other than classical releases) as "Jam Session at Commodore, No. 1." It was an auspicious beginning for the label and established the high standards that characterized all of Gabler's Commodore recordings. The music was quite different from anything being recorded at the time. It evoked the spirit of the previous decade, at the same time sounding fresh, vital, full of energy and excitement. Gabler's care with the recording process paid off as well. The intricate interweaving of the melodic lines and rhythms could be heard clearly. Pianist Joe Bushkin recalled:

> "That whole group that recorded for Commodore was like a fraternity. We all had a similar feeling about the music. We kind of considered ourselves to be the real jazz musicians, because we didn't play in the big bands, although actually most of us did, one time or another, except for Pee Wee. He was the real jazz musician . . . Pee Wee was kind of rattled but always right.

Pee Wee pulled out all the stops on that first session. "Love Is Just Around the Corner" gave him his first opportunity on record to stretch out for two full choruses. "In three minutes," wrote critic John McDonough,

> Russell has explored the clarinet from top to bottom, run through four or five drastic shifts of tone, created a couple of entrancing tunes and combined all this into a solo that could stand alone yet fits perfectly into what everybody else was doing. Young clarinetists from coast to coast memorized the solo after the record came out, but nobody ever improved on it.

An alternate take shows that Pee Wee's creativity, if not actually improving on the originally issued version, could produce yet another masterpiece.

The recordings and magazine articles helped establish a national reputation for Hackett's band. On February 5, 1938, as guests on a coast-to-coast radio program, Saturday Night Swing Session, they played "At the Jazz Band Ball" and "That's A Plenty." Unfortunately, the music was not preserved. A few days later, WMCA began broadcasting regularly from Nick's. A written account was preserved of the broadcast from March 9, and represents an example of the type of program to be heard at the club: "I Never Knew," "Time on My Hands" with a vocal by Red McKenzie; "Muskrat Ramble," "You Took Advantage of Me" with a vocal by the band's female singer, Lola Bard; "Indiana," "September in the Rain," "The Song Is Ended" and a composition credited to Pee Wee, "Village Strut" (undoubtedly a blues).

A couple weeks later, John Hammond again used Pee Wee on a recording session with Teddy Wilson. This time the results were issued immediately, and show Pee Wee fitting in quite well in a swing, rather than dixieland, context.

Hackett landed a recording contract with Vocalion for a session on February 16, 1938. On the first tune Pee Wee took a solo on tenor sax, his last recorded solo on any instrument other than the clarinet. The conception and execution of his tenor solo on "You, You and Especially You" sound relatively conservative compared with the clarinet solo he takes later in the same number.

Sales of Vocalion and Commodore records showed there was an audience beyond the confines of Nick's for unadulterated jazz. (The purer-than-thou attitude came back to haunt the Nicksielanders in the mid-1940s with the arrival on the scene of Bunk Johnson and other older jazz players whose fans regarded the music played at Nick's as commercialized.) Small-group jazz was beginning to regain some of the popularity it had lost to the big swing bands. Goodman's extremely popular swing ensemble belonged more to the "hot" dance band tradition than to jazz, argued the keepers of the flame around the bar at Nick's, but the "real" improvised jazz had created its own audience right there every night. Jam sessions organized by Gabler, Berton and others were attracting ever larger crowds. Trying to get more publicity for the fledgling movement, Ernest Anderson and Eddie Condon's future brother-in-law, Paul Smith, decided to lure Madison Avenue advertising executives to the jam sessions. According to Condon:

> In the spring of 1938, they came up with the idea of putting on jazz concerts late Friday afternoon in the Madison Avenue area . . . They finally made a deal with the Park Lane Hotel to rent their unused ballroom. I rounded up twenty-two guys and

things went well for many months. At one concert, however, a senior director of the New York Central, the company that owned the Park Lane, came in for a drink and he couldn't get near the bar. He was not used to such merriment on the staid old Park Lane premises and he was shellshocked. That ended our very successful run at the hotel.

The gang resumed the series at the Belmont Plaza, "but the atmosphere was different and we only played a few more Fridays."

The Park Lane series lasted into the early months of 1939 and were successful in attracting the kind of publicity Anderson and Smith wanted, although the manager who approved bringing the barbarian hordes into the hallowed ballroom was fired and the musicians shown the door. Among the journalists who became fans were members of the *Life-Time-Fortune* staff, including the editor of *Life*, Alexander King.

Speaking of Condon, Milt Gabler said, "Eddie was one of the most hustling, promotion-minded men I ever met in my life. And a delightful man. Every writer loved him. Newspaper people adored him. He rushed up to *Life* magazine and he said, 'Teagarden's in town.' [The trombonist was with the Paul Whiteman band at the time.] 'If I can get Milt to do another date, will you come and take a picture of it?'" King, a regular at the Commodore Music Shop and Nick's, liked the idea and agreed to it. Pee Wee's banjo-playing buddy, Charlie Peterson, who had become of one *Life*'s top photographers, was given the assignment. The idea at *Life* had grown to an eleven-page spread, encompassing the entire jazz scene and tracing its history up to that moment when, as *Life*'s headline stated, "Swing: The Hottest and Best Kind of Jazz Reaches Its Golden Age."

The Commodore-*Life* session took place on April 30, 1938, in the Brunswick studios with the same basic band as on the previous session, with the substitution of Teagarden for Brunis. Pee Wee was 20 minutes late getting to the session because Lola had hocked his clarinet for eight dollars. Pee Wee had been able to reclaim possession of his instrument only with the understanding that the creditors would be paid immediately after the session. They came along just to make sure and waited outside the studio while the *Life* people and the recording engineers busied themselves.

Life was delighted with the photographs. In the section "Hot Players Make 12-in. Records," the copy read:

> The most exciting swing performances have been given by groups of pick-up musicians who met in jam sessions or recording studios for the simple delight of playing as they pleased. Bix Beiderbecke always dreamed of getting together a great pick-up band, making twelve-inch records—long enough to give soloists a chance to round out their work. For the frustrated Bix, the dream never came true. But a short while ago in a Manhattan recording studio, the kind of band Bix longed for came together to make the twelve-inch records he wanted to make. They were from five different bands. All but two had played with Bix . . .

The photo layout included a picture of Bud Freeman; a group shot depicting Teagarden "sketching" one of the tunes; Condon and Gabler; Gabler bending over the cutting lathe; and one of Pee Wee playing an obbligato while Teagarden sings "Serenade to a Shylock." Next was a full-page shot of Pee Wee "in full croak," as one writer put it. The caption reads: "When a hot player takes his solo, he becomes the picture of complete and agonized absorption. This is 'Pee-Wee' Russell, who has a long elastic face. From his curved fingers to his arched eyebrows, Pee-Wee puts everything into the beautifully-wrought improvisations that come out of his clarinet." Peterson's photo of Pee Wee had been intended for the cover, but at the last minute it was pulled due to a late breaking news story: the abdication of King Edward VIII. The contents of the magazine had to be shuffled as well, and the story was pulled, appearing four months later in the August 8, 1938, issue.

Although missing publicity's brass ring—the cover of *Life*—

Pee Wee was catapulted to national fame through the article, and especially through Peterson's evocative photographs. The article also put Gabler's Commodore Records on the map. Condon had pulled off a coup. The records themselves were sensational. Bobby Hackett sculpted one of his enduring solos on "Embraceable You." The recording had an enormous impact on cornet and trumpet players at the time. Teagarden, on furlough from the Paul Whiteman "army" in which he served for five years, played throughout the session with vigor and imagination, and was featured on "Diane." The special chemistry between Teagarden and Pee Wee was evident throughout the session, perhaps nowhere more than during the trombonist's vocal on the studio-created blues, "Serenade to a Shylock," named in honor of the loan sharks waiting outside the studio door for Pee Wee. Pee Wee's contribution to the record begins immediately after the piano introduction with a heart-wrenching bent note wail that dominates the opening ensemble. He continues with a magnificent obbligato behind Teagarden's vocal. Elegant statements by Teagarden, Hackett and Bud Freeman follow; then the tempo doubles and Pee Wee plays another perfect solo and takes a break at the end in his patented style. The alternate take is just as good. Gabler paid the musicians at the end of the session, and Pee Wee paid the loan sharks. "Pee Wee imagined there were always gangsters after him," said Phyllis Condon.

> He was involved with loan sharks, and he got beaten up once. He got a little paranoid about gangsters later on. He told such horrendous tales to the boys [but] the danger wasn't anything what he imagined it to be. He lived day to day financially. He didn't have the relative stability of working in a big band.

Pee Wee may have become famous, but the pay at Nick's did not reflect his star status: $25 a night, and a good portion of that was consumed at Nick's bar and at Julius'. When money he earned on other gigs—a party or a record session—was added in, Pee Wee's weekly take probably averaged $200 to $250 a week at a time when an average weekly wage was $40 to $50. But Pee Wee had no regard for money; what he didn't spend, he lost or gave away. It went as fast as he made it. "That was Pee Wee," Milt Gabler said. "Never had any money. He lived from day to day. Every time I'd see him, he wants to borrow a little money. He said he'd work it off on Sunday at Ryans. It was just ten or twenty bucks. But when I'd do the concert, I'd pay him anyway."

Hackett and Pee Wee were featured at a "Carnival of Swing" concert at the Randalls Island stadium during the summer. "The carnival drew 24,000 cats," reported *Down Beat,* "and 3,000 fans stampeded the field in an effort to get near the bandstand during Duke Ellington's rendition of 'Diminuendo and Crescendo in Blue.'" (The same selection would turn Ellington's career around at the Newport Festival eighteen years later.) The article noted that Hackett's band played "Singin' the Blues" and "At the Jazz Band Ball," and "surprised the jitterbugs."

On June 25, 1938, the "Nicksieland" band, as it had become known, was featured on the Saturday Night Swing Session's second anniversary broadcast. They played "At the Jazz Band Ball," an Original Dixieland Jazz Band tune, again. Around this time, the band also appeared in a movie short called "The Saturday Night Swing Club," which purported to re-create for the cameras a typical radio broadcast. Musical film shorts were a popular feature in movie palaces at the time and were shown before or between features to fill out an evening's program. It was the only way most of the record and radio audiences could see how their favorite stars appeared in performance.

Bud Freeman was not yet a member of the Nicksieland crew. He had recorded several trio sessions for Gabler, but was still a featured soloist with Benny Goodman's band. In a well-publicized switch, Goodman had "stolen" Freeman from Tommy Dorsey's big band in March. Gabler hoped to capitalize on the publicity by listing Freeman as the leader of his next session on July 12. An original "Chicagoan" and an original tenor-sax stylist, Freeman was also a genuine improviser, as alternate record takes from this session reveal. Although a casual listener may mistake Freeman's well-defined style as repetitious, it is highly inventive within its framework.

On August 17, 1938, a national radio audience heard the Hackett band when they appeared on Paul Whiteman's network Chesterfield Show, during which the band gave forth with a tender version of Hackett's feature, "Embraceable You," and a stomping "Muskrat Ramble."

When the *Life* magazine article finally appeared in the August 8 issue, Pee Wee found he had become famous. Steve Smith, who, like Gabler, owned a jazz record store in Manhattan and who had started a collectors' re-issue label called Hot Record Society, felt Pee Wee's sudden fame warranted a recording session under Pee Wee's name. On August 31, 1938, after sixteen years as a professional musician—half of his life to that point—Pee Wee had his own session. It is doubtful if the selection of personnel was all Pee Wee's, although it certainly was a much more appropriate setting for him than usual, showcasing him in a more "modern" context. Instead of the usual "Nicksieland" combination, the band included swing musicians: trombonist Dickie Wells and guitarist Freddie Green, both from the Basie band, and Al Gold, a rising tenor star on The Street at the time. Also on hand were some old friends: Max Kaminsky, James P. Johnson, Duke Ellington's New Orleans-born bass player Wellman Braud, and drummer Zutty Singleton. In addition to four titles cut by the band, Steve Smith also recorded two tunes with Pee Wee in a trio setting accompanied by Johnson and Singleton. The records received rave reviews. *Down Beat* said the records "put to shame all other hot records which have been privately waxed recently, and have set a standard which will be hard to equal or excel." Smith was overjoyed with the brisk sales. But, due largely to Pee Wee's lack of interest in being a leader, it would be six years before another session would be recorded under his name.

Life was pleased with the reaction to its article on jazz and planned a follow-up, one of their "Life Goes to a Party" layouts. The party was staged in August in honor of a newly arrived critic from Indonesia, via Chicago, Harry Lim, and was held in the studio apartment of Burris Jenkins, a political cartoonist for Hearst newspapers. Lim had produced jam sessions in Chicago before coming to New York. Some of them predated Gabler's. Most of the prominent jazz musicians in New York, including Duke Ellington, Cab Calloway, Billie Holiday, Rex Stewart, J. C. Higginbotham, and the contingent from Nick's—Condon, Gowans, Kaminsky and Pee Wee—took part in the party for Lim.

The band at Nick's had been enlarged in an attempt to attract swing fans. Hackett was in charge. Arrangements by Hackett, by Brad Gowans and later by Buck Ram became part of the band's repertoire. The band alternated with Sidney Bechet's quartet. Vic Lewis, an English guitarist, made a pilgrimage to Nick's in 1938, and sat in with the band. The arrangement was fine with Condon, who was spending most of his time talking with the customers, so Vic took over the guitar chair "for the experience" for a few weeks. Soon, he arranged with the boys to fulfill one of his most cherished dreams: to play at a private recording session with them. Lewis rented the Baldwin studios, and on October 19 the musicians he had collected assembled there and cut six very relaxed sides, playing just for themselves. "It was amazing how the boys all gathered around to hear the playbacks and kept remarking on Pee Wee's work," Lewis said. "Everybody thinks he IS jazz with his solos drawing nothing but compliments." One of the tunes, "Tiger Rag," was played by the gang in a deliberately "corny" style, much as Pee Wee had been made to do on the "Whoopee Makers" session with Jack Teagarden in 1929, but this time it was strictly for laughs.

A couple of weeks later, Pee Wee participated in one of the huge public jam sessions that by that time were ubiquitous, this one beamed via short wave to England from the St. Regis Hotel roof and announced by a very enthusiastic Alistair Cooke. The *Life* spread was causing ever widening ripples for Pee Wee's career. Talk swirled around the tables at Nick's about a "history of jazz" Hollywood epic, starring the gang on the bandstand, a chance to tell the true story of jazz at last. Great projects were launched at Nick's nightly, most of them forgotten by morning.

One project that did pan out was when Ernie Anderson managed to get Pee Wee away from Nick's for a few weeks to front a band at Kelly's Stable on 52nd Street. Being a bandleader was a new experience for Pee Wee. A few weeks later, in November, 1938, Anderson and Condon got Pee Wee to front a band for a three-week stint at the Little Club, down the block from Kelly's Stable. That band included Max Kaminsky. One of the visitors to the club was the French jazz critic Hughes Panassie, who was producing a series of jazz records for Bluebird. Panassie enjoyed the group but didn't record them; he was in New York to capture what he thought was the "real" New Orleans jazz of Mezz Mezzrow and Tommy Ladnier. Kaminsky wrote in his autobiography, *My Life In Jazz*:

> The [Little] Club had become more of an entertaining night club, with singers, strippers, comedians and dancers and we played for all the acts. At a rehearsal one afternoon, one of the dancers handed Pee Wee, who was heading our four-piece band, a special arrangement for a forty-piece orchestra. It must have cost the poor girl a couple hundred dollars to have the arrangement made, but it was almost worth it to see the expression on Pee Wee—who was trying to play, read and direct all at the same time—when we had to sit there and silently count bars in the places where the French horns and flutes and violins were supposed to play. I finally told the poor kid we'd play 'Limehouse Blues' and she could make up a dance as we went along.

Down Beat alluded to underground problems in its notice of the band.

> Among the maestri who premiered . . . was Pee Wee Russell, who by this time has probably realized that one picture in *Life* does not make a band . . . Russell had a promising band but he worked under a handicap at the Little Club. Main trouble, it is alleged, was the lack of proper pay-offs, which is not at all inspiring to good swing. After three weeks the band exited . . . Russell will engage in several recording dates until he can reset his ensemble for a real chance.

Later that month, however, Pee Wee was back at Nick's with Hackett's band. They alternated sets with the Spirits of Rhythm.

The publicity from the *Life* spread took an unexpected turn when the Conn clarinet manufacturers approached Pee Wee to endorse their product. The endorsement surprised all who knew Pee Wee, since his instrument was always in a state of disrepair, held together by rubber bands and string. Full-page ads soon appeared in *Down Beat* and other music magazines featuring photographs of Pee Wee and Paul Whiteman's Chester Hazlett. The copy ran:

> Charles "Pee Wee" Russell: Swing Clarinet Sensation of the Day. For many years, Russell has been known as a fine swing artist. He played with "Bix" Beiderbecke, founder and idol of the modern "swing" movement, and is a leading exponent of the "Bix" school. Featured for several years at the "Famous Door." Played with Bobby Hackett's Swing Band at Nick's Cafe in Greenwich Village. Now fronting his own band at the Little Club—52nd Street, New York City—popular rendezvous for musicians and swing lovers of the metropolis—the shrine of the old "Bix" tradition.
>
> Chosen as the outstanding swing clarinetist for the Jitterbug picture soon to be filmed. Also selected by the Hot Record Society to carry out "Bix's" dream of making hot records which would stand for all time as classics of jazz. Critics rate the new re-

cordings by the "Pee Wee" Russell Recording Unit as absolute tops in "hot" swing and hail "Pee Wee" as the greatest swing clarinetist of modern times.

If Pee Wee's friends, who knew the condition of his clarinet, were mystified, the reaction to the ad by the company's salesmen was outright hostility. One music-store owner showed a Conn salesman a copy of the new catalogue, and turned to the particularly dissipated-looking photo of Pee Wee that accompanied the article. "Good God, have they got *him* advertising *our* clarinets?" gasped the salesman. "Anytime he'd sell a clarinet I'll eat it. I could make a better noise with a lead pipe under water."

Ernest Anderson's media contacts, especially in the advertising business, led to Pee Wee and Condon appearing in a couple of other advertisements. One of them shot by Charles Peterson, was for a Kellogg's cereal. The full page spread showed Condon with a trumpet tucked under his arm, extolling the laxative properties of the cereal to a constipated Pee Wee, seated behind a drum set. The layout ran in several Canadian newspapers in 1940.

On the advice of an agent, Sidney Mills, from the Music Corporation of America, Bobby Hackett put together a big band. In April, *Down Beat* reported the band was formed "after leaving Nick's due to a disagreement with the management and was scheduled to break in with a series of one-nighters in New England." Even though Pee Wee had sworn never to work in a big band again after leaving Prima, he couldn't say no to Bobby. One of their first engagements was at the New York's World Fair. With a few personnel changes, the band recorded for Vocalion on April 13, 1939. The recordings would be among the last to display Pee Wee in the big band setting until the very end of his career. During the last week of the month, the band opened at the Ben Franklin Hotel in Philadelphia. Ernest Anderson remembered:

This was a very, very special date, highly sought after by all the bands because they had twelve remote broadcasts a week, and bands really lived on those broadcasts. A band would go in to a hotel at minimun union scale and then with the broadcasts they'd go on the road for one-nighters and clean up.

They were all supposed to leave for Philadelphia at 10 a.m., and they were to meet at a musicians saloon across the street from Columbia records on Seventh Avenue and 52nd Street. Pee Wee and Lola were living together at that time—this wasn't a continuous thing by any means. They'd get together, break up, get together and so forth. But at this time they were living at a place called Center Rooms which was between Broadway and Seventh Avenue about 48th Street. It was the lowest-down hotel in New York City. No question about it.

Everyone had to have a tuxedo to play at the Ben Franklin and when Pee Wee came home after playing until 4 a.m. that morning to get his tux, there was a sign on the mirror, written in lipstick which said, "You'll find your tuxedo hanging in the closet." He went to the closet and it was hanging there in half-inch strips. She had taken a razor blade and just cut it into strips. So he arrived at the saloon with no tuxedo.

It turned out that Bobby Hackett had bought a new tuxedo for the engagement. Now, Bobby was at least a foot shorter than Pee Wee, but they sent somebody in a taxi and got Bobby's old tux and that's what Pee Wee opened in that night, with his arms sticking about a foot out of the sleeves. That was my last sight of Lola.

Two weeks later, the band was spotted by *Metronome* at the State Ballroom in Boston. In May, the band made a special Monday night appearance at the Famous Door in New York. By the end of May, the band was back at the Ben Franklin in Philadelphia. Claire Martin was the vocalist. The band scuffled through a series of one nighters in June and the beginning of July. By that time, it was beginning to fall apart. Discipline was a problem: they had an opportunity to audition for a regular radio program, but it was cancelled when too many members of the band failed to show up at the appointed hour. The band was underfinanced; Hackett, already $2,800 in debt to the agency, was ready to call it quits. Pee Wee, Condon and Jacobs decided it was time to leave. The three briefly played with a band put together by pianist Artie Schutt for a job at O'Leary's Barn in New York. Then, in July, Pee Wee joined a new band Bud Freeman was fronting at Nick's. Max Kaminsky remembered that the idea for the band had been discussed a couple years earlier when

> Eddie [Condon] called me to come to town to play the Princeton alumni reunion with Bud, Brad [Gowans], Pee Wee, Dave Bowman and Dave Tough. After the job, we had a long talk about the possibility of staying together and trying to make a go of it with our own band. I'd join only on the condition that it be a cooperative organization . . . This suggestion met with unanimous approval and a five-year charter for the Summa Cum Laude band was drawn up.

"Eddie Condon had a band at Nick's, Freeman said, "and when I left Benny Goodman I made up my mind I would never play in a big band unless it was my own. I hadn't the faintest idea what I was going to do except that I wanted to play my own music and be a jazz soloist." In April, 1939, Freeman got a call from Ernie Anderson.

> He asked me if I'd like to front what was really Eddie's band. With me leaving Tommy Dorsey to join Goodman, they were sort of feuding over me and I'd got a lot of publicity—*Down Beat* and all these papers built it up. Lot of rubbish, really.

Freeman agreed to front the band

> with one provision: that I'm not held responsible for anything people do. I just want to play. We opened at Nick's and you couldn't get in, it was such a smash. Several recording companies came down and we took the best one, which was Victor, and we made those sessions for Bluebird.

Phyllis Condon dubbed the band Bud Freeman's Summa Cum Laude Orchestra. Freeman called rehearsals three times a week to work up new arrangements. Brad Gowans wrote the ideas down, and the band began to develop its own sound. The jazz press had fun with the band's name, referring to them as the "Come Louders." Condon referred to them as the "Come Over Loaded" band.

The first recording session for the Summa Cum Laude orchestra was for Bluebird on July 19, 1939. Included were two compositions by Freeman: the band's theme song, "Easy To Get," and "The Eel." The other two were old chestnuts (even then), "I've Found a New Baby" and "China Boy." Pee Wee's playing throughout is superb. The records sold well, but the band switched to Decca for further recordings.

Before the Summa Cum Laude band's first Decca session, however, a Yale student, George Avakian, approached Jack Kapp, Decca's owner, with a unique idea: an album of jazz records commemorating the famous recordings made by McKenzie and Condon's Chicagoans for the Okeh label twelve years before. The idea was to re-create the band as closely as possible and, by using a couple of other "Chicago" bands, to produce a "state of the art" jazz album. It would be the first time jazz records had been issued in an album, a format previously reserved for classical works and show albums. There had been a couple of reissue albums, notably a Bix Beiderbecke memorial album on Victor, but this would be the first time new jazz recordings would be released in album form. To Avakian's surprise, Kapp said yes. The only artists who appeared on both the original session and Avakian's were Freeman, Joe Sullivan and Condon. Pee Wee took the role that had been played by Frank Teschemacher.

"One take of 'Someday Sweetheart' was just a rehearsal, but a wonderful one," Avakian said. "It ran too long. It ran out during the last chorus. I hated to lose any of the solos so Eddie said why don't we just play the first chorus like this—he indi-

cated a shortened chorus. That's why you don't have a first ensemble chorus on it." The recording process at that time did not
allow the musicians to immediately hear what they had recorded.
The music was inscribed onto wax discs which later were plated
to produce the metal stampers used to manufacture test pressings. It might be a week or more after the session when the
musicians would be able to hear the results. "The fact that you
couldn't play back meant you had to be conscious of quality the
first time around." Avakian said. "You didn't have this attitude
of, 'Oh well, we'll just play it and you edit it later." The quality
of the music was sterling; the album was successful musically
and financially and launched Avakian on a long career as one of
the most sought-after "artists & repertoire" men in the business.

The "Come Louders" continued to record for a variety of
labels both on their own and providing accompaniment for several singers, including Lee Wiley on the Liberty Music Shop
label, Buddy Clark on Variety, Teddy Grace on Decca, and
another John Hammond discovery, Doris Rhodes, on Columbia.
For Decca, the band recorded an album of tunes associated with
Bix Beiderbecke's first band, Wolverine Orchestra.

Gabler's Commodore label was still very active. On November 30, 1939, he recorded Kaminsky, Gowans and Pee Wee with
a different rhythm section under Condon's name. Four remarkable performances were issued, including a classic, "I Ain't
Gonna Give Nobody None of My Jelly Roll," which Pee Wee
always proudly noted was in the Library of Congress as an outstanding example of jazz clarinet, and "Ballin' the Jack," in
which Pee Wee plays tricks with the rhythm that seem to defy
the beat, yet swings mightily throughout.

The usual Greenwich Village crowd continued to use Nick's
as a local meeting place. Condon was always working the audience, trying to further the interests of the band. *Down Beat*
announced in November: "Condon To Be a Thespian." Dorothy
Baker's book, *Young Man with a Horn,* supposedly based on the
life of Bix Beiderbecke, had reached the bestseller lists, and

Vinton Freedley had adapted it as a play. Condon was to act in the play, and Burgess Meredith was selected for the leading role. Bud Freeman's band was to play from the pit. Apparently nothing ever came of this scheme.

In November, Nick decided if one band was so successful, why not bring in a second. Muggsy Spanier had put together a fine small group, the Ragtimers, at the Hotel Sherman in Chicago and had recorded some excellent sides for Bluebird. Nick brought them in during November, alternating with Freeman's band. Muggsy's band was something different for the crowd at Nick's, and, according to several witness proved even more popular than the Summa Cum Lauders. According to an article in *Jazz Information,* Pee Wee was supposed to have played on the Ragtimers' first New York recording session on November 10, but he did not show up and his place was taken by the band's regular clarinetist, Rod Cless. A couple of weeks later, Nick gave Freeman his notice, just in time for the band to take part in one of the greatest all-star flops ever staged.

"Swinging the Dream," a jazz version of William Shakespeare's *Midsummer Night's Dream,"* and featuring a predominantly black cast, was written by Gilbert Seldes (a regular at Nick's) and Erick Charrell. The score was composed by Jimmy Van Heusen, who borrowed themes by Felix Mendelssohn, and lyrics were written by Eddie De Lange. The cast included Louis Armstrong in the role of Bottom, Butterfly McQueen as Puck and Maxine Sullivan as Titania. Dorothy Dandridge appeared with her sisters, Vivian and Etta, as pixies. Don Voorhees conducted the pit band; James P. Johnson was the rehearsal pianist; and there were two small bands that appeared in boxes at each side of the stage. One was Benny Goodman's sextet with Charlie Christian. The other was supposed to be the John Kirby group, but that was changed at the last minute to Bud Freeman and his Summa Cum Laude Orchestra with Zutty Singleton on drums.

On November 25, 1939, the first of three preview performances was given. The show opened at Radio City's Center

Theatre on November 29. The Summa Cum Laude band was cut to two numbers, "preludes" to the first and second acts. The reviews were not kind: "An orgy of wasted talent," wrote *Billboard*. The show closed after 13 performances on December 9. The first week's gross had totaled only $12,000. "It was just kind of stupid," Freeman said. "If it had just been a revue it would still be running."

Following the demise of the show, the Summa Cum Lauders went into the Brick Club on 47th street, replacing Hot Lips Page's band. Although the club was "a real dive, sort of an underworld rendezvous," according to Freeman, it had a radio hookup and broadcast every night. The musicians were paid $37 a week. The personnel was stable except for the drummers. Danny Alvin, Al Sidell, Morey Feld, George Wettling and Dave Tough each played at various times. The band struggled through the new year, 1940. The Cum Lauders, with Dave Tough on drums, were set to open at the 711 Club at 55th Street, east of Fifth Avenue for two weeks, but the date was cancelled at the last minute. Tough had left Jack Teagarden's band to rejoin the band for the job. The band finally opened at Kelly's Stable on January 26, 1940, with Tough, and Pat Peterson on bass. Gowans was playing with Joe Marsala at the Fiesta when the gig began, but he was soon able to rejoin the band. *Jazz Information* reported that the "Freeman group will hold jam sessions every Sunday afternoon from 5 to 7, with various musicians present as guest stars. Other plans now being made have Eddie Condon leading special relief bands on Tuesdays, the band's night off."

On February 6, 1940, Pee Wee participated in a special concert for children at the prestigious "experimental" school, the Little Red Schoolhouse. The session had been set up by Helen "Daisy" Decker, a childhood friend of Pee Wee's who was working as the photo editor at *The Ladies' Home Journal*. Charles Peterson was recruited to take photos showing Pee Wee holding the tots' attention, Peterson's young son Don among them.

Pee Wee continued to be very active in jam sessions and private parties, usually as part of the band, but sometimes he had the opportunity to stretch out as the only horn. One notable occasion was in February when he played a private jam session at the Pleasantville, New York, home of Hubbell Young, an editor at the *Reader's Digest* and a habitué at Nick's. Pee Wee was accompanied by Joe Sullivan and Eddie Condon. His regular job continued to be with the Summa Cum Lauders. When the job at Kelly's Stable finally ended, the group went into the Brick Club, opening there on February 16, 1940. By March, Tough had been replaced by Morey Feld.

According to Max Kaminsky, singer Lee Wiley got the Freeman group their next job (booked by Willard Alexander), at the Panther Room of the Hotel Sherman in Chicago, where they opened at the beginning of May, 1940, with Fred Moynahan on drums. The musicians were paid $60 a week. Wiley sang with Freeman's group. Violinist Stuff Smith's band featuring Jonah Jones and Cozy Cole, alternated with Freeman. *Down Beat* interviewed Freeman when he arrived there and reported the band had an offer to play at the Copacabana in Rio de Janeiro during the summer. "We'll probably go there if we are not held over at the Sherman," Freeman said. "All the boys would like the trip—travel is so broadening! But first we'll have to see what happens in Chicago."

Max Kaminsky recalled:

> We did two broadcasts a night and we also played for the floor show, which included a clown who handed out balloons to the patrons and had them participating in kiddie cart races, and a Spanish dance team for whom we played bullfight music. There were the days when the publishers were still going around plugging new songs for the bands to play on the air. Brad would sit up all night writing melody arrangements of the songs: "April Played

the Fiddle," "I'll Never Smile Again," "The Wind and the Rain in Your Hair," and the next night we'd play the medleys mixed in with the jazz classics. In spite of everything, the band sounded fine and Lee was singing as only she could.

On Monday nights when the lounge was dark, the boys repaired to the estate of Edwin "Squirrel" Ashcraft in Evanston. The son of a wealthy Chicago lawyer, and quite a successful one himself, Squirrel had made a practice of entertaining jazzmen at his home. An amateur musician (piano and accordion) himself, having made some private recordings as part of the Princeton Triangle Club Jazz Band in the 1920s, Squirrel had loved the music ever since Bix Beiderbecke appeared on the Princeton campus to play at fraternity parties. Regular guests at Squirrel's included such former Chicago figures as Jimmy McPartland and Bud Freeman. During the Sherman Hotel engagement Pee Wee jammed at Squirrel's at least once. Gowans and Free. played there more often.

Although the Chicago jazz fans frequented the Panther Room, its business had fallen off by the end of the run on June 6. "When the Panther Room engagement ended," Kaminsky recalled, "there wasn't a soul around offering us another job. Eddie even resorted to sending a wire to Bing Crosby, but Crosby's return wire squashed that hope. Los Angeles was still a hillbilly town and didn't dig, he said." The job at the Copacabana also fell through. Ernest Anderson had arranged a tour of New England ballrooms for June, but the contract called for a big band. Several men were added to the Summa Cum Lauders for the job, but Pee Wee declined to join. One of the finest small bands of the swing era, "one of the most cohesive of its time," according to critic Richard Hadlock, had ceased to exist. Though the Summa Cum Lauders were no longer playing in public, they did some more recording. On July 14, 1940, they assembled in Columbia's studios for that label's response to the success of Avakian's Chicago jazz album on Decca. Brad

Gowans was replaced by Jack Teagarden. Released as Bud Freeman's Famous Chicagoans, the records were musically as successful as the Decca album. That the recordings made by the Summa Cum Laude orchestra were not commercially competitive with contemporary releases by Benny Goodman is irrelevant as far as the quality of jazz is concerned. The Summa Cum Laude sessions, and almost all of the other recordings made by Pee Wee during the late thirties and early forties, show him at the peak of his creative powers. As superb as the contributions of the others with whom he played were—Hackett and Freeman especially—it is Pee Wee's work that has best stood the test of time. With astonishing consistency he maintained a high level of creativity that continues to delight the listener.

8 • • • •

Town Hall
Concerts

*D*espite the lack of regular work for the Summa Cum Laude band, jobs were steady for Pee Wee. Extra gigs—record dates, private parties, public jam sessions—were plentiful. Over the next few years the jam sessions that Gabler and others had started in radio studios and empty clubs blossomed into full-scale concerts held in some of New York's most prestigious halls. Pee Wee found himself at the center of the activities.

Pee Wee and some of the other Cum Lauders were welcomed back at Nick's. By September, Bobby Hackett had taken over the band, adding two saxes and dropping Joe Sullivan and Pee Wee. The boys had been a trifle too bad for Nick once again. But the Summa Cum Lauders got together for one last record session, backing up singer Teddy Grace for Decca on September 26, 1940. Pee Wee did not have much of a chance to play on the session, being limited to a brief solo spot on "Sing—It's Good for Ya."

Sunday afternoon jam sessions had spread to venues all over the Northeast. A new audience was developing in response to being able to see and hear legendary performers from the golden years of jazz. At last jazz had found an audience that did not think of it as some kind of disreputable, bastardized dance music, but as art music. It was being taken seriously by a growing number of fans. The small jazz press was burgeoning with publications such as *Jazz Information* and *Jazz Report*, the later published, edited and largely written by the Chicago pianist Art Hodes. These publications, with scholarly articles about the history of the music, interviews with many of the early figures of jazz, and reviews of performances and recordings, were among the first to treat jazz as worthy of serious study. While *Metronome* and *Down Beat* also covered jazz extensively, their approach was more of a commercial nature—the business of jazz rather than the aesthetics.

Although the publications fanned the flames of popularity, jazz remained only slightly more than a cult phenomenon, a fad of sorts, during this period when the big swing bands were still in full glory. But the music had gained a new degree of respectability; the novelty of hearing jazz in a concert hall was wearing off. The crowds that came to hear the Condon groups, as they were becoming known due to the success of his Commodore records, were becoming better educated and more involved with the music itself.

In addition to all the dates he had in the city, Pee Wee was a frequent participant in sessions outside the metropolitan area. During the fall, for example, Pee Wee was scheduled for a series of Sunday jam sessions started at the Greenhaven Inn on the Boston Post road in Larchmont, New York.

Music publishers, Leo Feist, Inc., approached Pee Wee for their folio series, *All-Star Series of Modern Rhythm Choruses*. As the ad explained: "Each book is individually arranged by a Top Musician and contains his versions of 'Sleepy Time Gal,' 'My Blue Heaven,' 'Swingin' Down the Lane,' 'In a Little Spanish Town,' 'The Darktown Strutters' Ball,' 'Linger Awhile,' 'At Sundown,' 'China Boy,' 'Sunday' and 'Ja-Da'." Twenty-eight artists, from Buster Bailey to Joe Venuti, were included in the series. Pee Wee's versions were published late in the year.

At the beginning of November, Pee Wee was again installed at Nick's, but this time his accompaniment was something very different from the sounds that had become known as "Nicksieland." In addition to the clarinetist, the quartet consisted of Tony Ambrose, the leader-accordionist; Bill Clifton, a fine young pianist, and bassist Fowler Hayes. The band alternated sets with the Hackett aggregation. Apparently, the accordion band was a penance Nick imposed as a condition of reemployment for Pee Wee.

Milt Gabler produced another Condon-Commodore session on November 11 that reunited Pee Wee with Fats Waller on perhaps their best session together. Because of Waller's exclusive recording contract with Victor, Gabler's label listed the pianist as "Maurice," Fats' son's name. The great chemistry between the two jazz men—last recorded on the Billy Banks Rhythmakers session more than nine years earlier—steams on all four titles.

Pee Wee played at the Greenhaven Inn on November 24, 1940, when he was featured with Hot Lips Page. He was scheduled to appear at the other sessions there, but other jobs prevented his participation. During this time, Pee Wee was the nominal leader of a trio at one of the most sophisticated night spots in New York, Billingsley's on Park Avenue. The group, named "The Three Deuces," at first included the Mound City Blue Blowers' original banjoist, Jack Bland, who had long since switched to guitar. By the time the engagement ended, the group consisted of Pee Wee, Joe Sullivan and Zutty Singleton.

Pee Wee next joined a group led by Bud Freeman, for a one-week stint at the West End Theatre in Harlem. The band then traveled to Miami Beach, where they caused a sensation at the Paddock Club at 7th Street and Washington Avenue for nearly a month. Pee Wee was back in New York in time to appear in the first of a series of weekly concerts produced by Ralph Berton,

WNYC radio's Metropolitan Reviewer. The event was held at the Village Vanguard on December 29, 1940, and also featured Coleman Hawkins and Lester Young. The session began at 2:30 and ended at 8 p.m., all for a 65-cent cover charge. After the first session, Pee Wee, Joe Bushkin and Condon participated in an informal jam session at the home of *Time*'s music critic, Carl Balliet.

Berton's next Village Vanguard session, on January 5, 1941, featured Pee Wee with Frankie Newton "and many more." Portions of the jam sessions were broadcast over WNYC and apparently were recorded on 16-inch acetate discs by the radio station, but none of them have been found.

Berton promoted Pee Wee's appearance on the ersatz Chamber Music Society of Lower Basin Street the following day. The popular radio program reached its audience by presenting jazz in a comic setting. The show had pompous "classical" music announcers introduce the groups and discuss jazz in a parody of academic analysis. Each program featured a guest artist or group, and on this occasion the group consisted of Pee Wee, Joe Sullivan, Eddie Condon and Dave Tough. It was the first time Pee Wee had been featured on the air as a soloist accompanied only by a rhythm section. They played two blues numbers. After the show, a 15-year-old clarinet student who had been in the audience approached his hero for some technical advice, according to Berton:

> "Mr. Russell . . . I've always w-w-wanted to ask you—how do you manage to g-get that wonderful *dirty tone* on your instrument? I've tried everything—I tried soaking my reeds in water overnight." Pee Wee, hunching and twisting as he finally glanced at his tremulous admirer, advised: "Why don't you try soaking your *head* in *whisky?*"

Pee Wee was also a regular at Gabler's jam sessions at Jimmy Ryans on Sunday afternoon. On January 19, he appeared there with Bobby Hackett, Brad Gowans, and Joe Sullivan, and on

January 26, 1941, Gabler used Pee Wee in one of his marathon jam sessions at the club.

In March, Nick agreed to rehire Pee Wee for Jimmy McPartland's band.

Gabler recorded Pee Wee for Commodore in a trio format on March 25, 1940, with Joe Sullivan and Zutty Singleton. The four titles were issued under the name of The Three Deuces. The records show Pee Wee at the height of his powers, "a kind of distillation of Pee Wee Russell at this point in his development as an artist," critic Dan Morgenstern has written. One of the titles was a blues, "The Last Time I Saw Chicago." Ironically, although the jazz press and critics pigeonholed him as a Chicagoan, the Windy City was Pee Wee's least favorite. "I've never had anything but bad luck there," he said. "That lousy city was always a jinx to me. Whenever I have to go to Chicago I always feel as though something bad is going to happen to me." Whatever the title meant to Pee Wee, his intense playing made it a classic. Critic John McDonough writes that Pee Wee's playing on the tune "must be considered among the greatest examples on record of blues playing," and "Russell, in one bone-chilling 24-bar solo, has made the most eloquent of statements."

After The Three Deuces recordings had been released, Pee Wee had a return engagement on the Chamber Music Society of Lower Basin Street broadcast. On December 3, 1941, the trio played two of the tunes they had recorded: "Deuces Wild" and another classic version of "The Last Time I Saw Chicago." Chicago may have been on Pee Wee's mind because McPartland had received an offer to play at the Brass Rail there and asked Pee Wee to be in the band. On July 6, 1940, McPartland opened there with Pee Wee—more nervous than usual at being back in Chicago. In McPartland's account:

> I had been working at Nick's when Fred Williamson, an agent, came in and had some good loot for a stay at the Brass Rail in the Loop. We go to work one night and there is no Pee Wee. Squirrel Ashcraft called and said Pee Wee was in jail for stealing milk.

Apparently, there was a milk wagon and Pee Wee and Sullivan were going home after work. It was about four in the morning, and Pee Wee snatches a quart of milk from the truck. The milkman sees it, calls the cop, and the next thing Pee Wee's behind bars.

Ashcraft recalled that "Condon's uncle, who was a power of some kind, interceded. So the judge just told Pee Wee he didn't want to hear of him drinking any more milk. I got him out of jail that time." The idea of the judge ordering Pee Wee to refrain from drinking milk was a standing joke among the musicians, but Pee Wee had been living on milk for years, frequently pinching bottles from doorsteps on his way home from work, a few steps behind the milkman. Milk was one of Pee Wee's few sources of nutrition.

McPartland's band had a successful month-long engagement in Chicago and was broadcast nightly over a local CBS radio hookup. Pee Wee's health was poor and he was unable to return to New York immediately after the gig was over. Pee Wee, reported *Down Beat,* is "recuperating in Chi with thermometer in hand before venturing more than a block from his room, but he intends to leave as soon as possible." Meanwhile, rumors circulated that McPartland and his group would be chosen to appear in an Orson Welles' film which was to be a history of jazz, using a documentary approach and starring Louis Armstrong. The project was cancelled before any commitments were finalized.

Pee Wee finally returned to New York by late 1941 to resume working at Nick's again, with cornetist Wild Bill Davison. It was Davison's first New York appearance. He had been known to the Chicagoans during the 1920s as a Bix-inspired player, but he had spent most of his career in the Milwaukee area, where he developed the very personal style described by his nickname. His was a forceful, fiery approach, quite unlike the playing of McPartland, Hackett or Spanier, and it made a powerful impression on the patrons at Nick's. For a number of years

Davison had been persona non grata with the Condon gang because in 1932 he was driving the car in which their hero Frank Teschemacher met his untimely death when barely 25 years old. The accident was not Davison's fault, but for many years the others felt uncomfortable having him around. His debut at Nick's marked a reunion and acceptance by both musicians and fans.

Pee Wee continued to be a mainstay of the many jam sessions that proliferated in New York. On November 9, 1941, he appeared at a Jimmy Ryan's jam. On the twenty-third, he was back at Ryans.

Milt Gabler's mailing list had grown through the years, and his sessions at Ryans were, for the most part, standing room only. The "official" band in early 1941 consisted of Bobby Hackett, Brad Gowans, Pee Wee, Condon, Dave Bowman, piano, Bill King, bass, and Zutty on the drums. But the vast array of extra talent available in New York always outnumbered the band. Joe Sullivan, Albert Nicholas, Sandy Williams, Marty Marsala, Teddy Bunn and his Spirits of Rhythm, Billie Holiday, and many others regularly participated. "There was a moment there," wrote Max Kaminsky, "at the Ryans sessions, when hot jazz seemed at its purest . . . At Ryans the music was the thing and when a musician was building a solo, you never heard a sound from the audience. You could *feel* them listening."

But before the end of the year, the scene changed abruptly as the United States was thrust into World War II. The sessions continued, but many of the fans and musicians were drafted. Pee Wee reportedly tried to enlist several times, to no avail. According to Jeff Atterton, "Condon once said they'll draft Pee Wee when the Japs get to Rockefeller Center."

In the early months of 1942, Pee Wee took special notice of an attractive, dark-haired young woman named Mary Chaloff who dropped into Nick's quite frequently. A sometimes-model, she was living the bohemian life in the Village. Although Mary frequently maintained that she didn't understand Pee Wee's

playing and felt "embarrassed" by it, the two soon became insep-
arable. Although she could be as fiery, outspoken and unpredictable as Lola had been, her relationship with Pee Wee was not a
destructive one; she was fiercely protective.

Mary was born on April 26, 1909, in New York, where her
mother's father, fleeing religious persecution in Odessa, Russia,
had settled. Once established in New York, the grandfather had
brought the rest of the family over one by one. Mary's mother,
née Clara Plotnikoff, and her father, Abraham Chaloff, arrived
shortly thereafter. It was an artistic and musical family: an
uncle, Eugene Plotnikoff, had been conductor of the Imperial
Opera in Moscow before the Revolution; later, he led the New
York City Symphony.

"I was born on the Lower East Side," Mary told Whitney
Balliett. "I was a charity case and the doctor who gave me my
name and signed the birth certificate was Dr. Edward Condon.
Isn't that weird?" She was the second youngest in a family of
three sisters and three brothers. Her brother Herman Chaloff
became a protégé of Aaron Copland, and a noted composer and
music educator. Her first marriage was at eighteen, but it didn't
last long. Her second marriage, also short-lived, was to an artist
who lived in the Village.

It wasn't long before Pee Wee moved in with Mary at her
apartment in a wooden building at 125 West 21st Street. The
building was described as a "flea-bag" and a "fire-trap" by its
occupants, who lived in a kind of happy communal squalor. Rent
was $21 per week "for a room that would go for eight in peace-
time," Pee Wee said. Nevertheless, it was a step up for him.
"She took care of Pee Wee," said Phyllis Condon of Mary. "He
would have been a derelict otherwise—probably the most famous derelict in the world, but still a derelict. He couldn't take
care of himself." Mary began accompanying Pee Wee to recording sessions.

On January 28, 1942, Gabler assembled the band for another
Commodore session. Pee Wee later named one of the tunes cut

at the session, "Mammy O' Mine," as his least favorite recording. Another selection, a blues, was entitled "Tortilla B Flat" a
pun on the title of the latest book by one of the Nick's regulars,
John Steinbeck.

Ernie Anderson and others felt the success of Gabler's and
Berton's jam sessions had grown to the point where it would be
economically feasible to present the music in the more formal
setting of the concert hall. "Ernie convinced me," wrote Eddie
Condon, "that it was time to put real jazz in a concert
hall . . . together we persuaded Fats Waller to be our first
attraction." Anderson booked Carnegie Hall for January 14,
1942, and backed the great pianist with a band drawn primarily
from Nick's. The concert was a financial success, if not an
artistic one. Backstage Fats imbibed a bit too much and was
disoriented during the concert, playing what sounded like the
same tune over and over again. The reviews were not flattering.
Waller looked "unnatural" in tails, opined *Down Beat*.

> At the very close a group of alleged 'Chicago' musicians compris-
> ing Eddie Condon, Gene Krupa, Pee Wee Russell, Max Ka-
> minsky, Bud Freeman and John Kirby (Huh?) ran onto the plat-
> form to produce an anemic, uninteresting session which was
> spotlighted by the rendition of the "Star Spangled Banner," out
> of tune, and in some spots, with the melody missing. Some of the
> boys, it developed, didn't know the song.

Legend has it that the musician who didn't know the national
anthem was Pee Wee. But the concert was important in proving
Anderson right: people would come to staid Carnegie Hall to
hear the "barefoot gang."

On February 21, Anderson put on another concert, this time
at Town Hall, 123 West 43rd Street. It was advertised as "A
Chiaroscuro Jazz Concert under the direction of Eddie Condon"—chiaroscuro because the events began in the twilight at
5:30 p.m. and ended after nightfall at 7 p.m. Ticket prices for
the concert ranged from 75 cents to $1.50. It was the first in a
series of four concerts, the others with similar personnel con-

ducted on March 7, March 21 and April 11. From then on, Anderson and Condon, and then Condon on his own, would present concerts in Town Hall and other New York venues through the 1960s.

At first, low attendance almost ended the series. "It was a hit in an empty hall," Anderson said. "The critics wrote extravagantly about it, but there was nobody there." The first four concerts were recorded in their entirety by the Rockefeller Committee, which was attempting to organize a network of educational and cultural radio stations throughout Latin America. Pressings of the concerts, with dubbed announcements in Spanish and Portuguese, were to be sent to 500 stations in thirty countries. An unedited set was also presented to the Library of Congress. The radio network project, however, was an early casualty of the war, and it is doubtful if the pressings were ever made, let alone distributed.

Anderson had another promotional project up his sleeve: a shot at a brand new medium. The Condon band was the first jazz group to appear on American television, in 1942. The April 16 broadcast was transmitted to approximately 350 receiving sets in New York from the CBS experimental studios above Grand Central Station. Gilbert Seldes had just been appointed program director of the fledgling operation, when his assistant, playright Worthington Minor, happened to walk past Town Hall when one of the concerts was taking place. "Mr. Minor wasn't really interested in jazz," Ernest Anderson recalled,

> but he knew that his boss, Seldes, was. He had already decided that musicians playing didn't make good TV because their eyes were always anchored to the written scores. But these musicians didn't have any written music. They were expressing every nuance of the music with their faces and their bodies.
>
> Monday morning I had a phone call from Gilbert Seldes. In the end he gave us two half hour programs a week, one was always a pianist with a drummer, the other was a little band with Eddie, usually seven musicians. I clearly remember two shows.

One was Harry The Hipster accompanied by Sidney Catlett. Harry, who went under the name of Harry Gibson although he was baptised Harry Raab, was an Eddie Condon discovery. The other TV show was Eddie's small band. Woody Herman's band was headlining at the Paramount Theatre. We got Woody's schedule, talked to him and he dashed over between shows and sat in. Even Woody got paid union scale by CBS-TV. Seldes was delighted with the show. He became a close friend and wrote program notes for subsequent Eddie Condon concerts.

"Condon Clicks on Television," *Down Beat* headlined. "The television concert Eddie Condon and Ernie Anderson tossed over CBS last month was such a success that they've been doing Monday night shots ever since with a drummer and pianist every time.

"After about a month of these shows," Anderson said, "the war put an end to them. The CBS-TV system was requisitioned by the Defense Department and all the telecasts from that moment on were war-related instruction, such as how to roll bandages." But Anderson and Condon had seen the potential of the medium long before most Americans were even aware of it.

Another special date was a benefit at the "progressive" Walt Whitman School in New York in which the guest of honor was Louis Armstrong. Louis jammed with the Condon band, but the trumpeter drew the line at singing the blues because, he explained, the only ones he could remember were dirty and not fit for the kids. For more than an hour, the band thrilled the students and an overflow crowd of adults as well.

Pee Wee continued to perform nightly at Nick's and during May the band was under his "leadership." It included Wild Bill Davison and Georg Brunis. In July, Wild Bill put together a band which included Pee Wee for a month-long job in Chicago's Capitol Lounge. From there the band was to play the Ken Club in Boston, but cancelled at the last minute. George Frazier reported in the *Boston Herald*:

Pee Wee Russell didn't show at the Ken, but Bill Davison is the man to blame. Davison called from Chicago and said the local wouldn't let him out of his contract to play at the Capitol Lounge.

Finally, on August 13, Frazier reported that the band

pulled in from Chicago around 6:30 Saturday night, a bit bleary-eyed and lacking their customary disposition for the remark witty, and they rushed up onto the bandstand at the Ken and began to play. And that was it!

In a reivew four days later, Frazier wrote,

The Davison band (although it needs a trombone and bass desperately) is a lovely affair. Pee Wee Russell is wonderful, of course.

Another season of Town Hall concerts began on November 7, 1942. By this time, the format had developed into a complete show. This one opened with a "Memories of Chicago" segment. Bobby Hackett was featured on "Ja Da" and Pee Wee on "Love Is Just Around the Corner." Harry Gibson, a fresh-faced Juilliard piano student, played Bix Beiderbecke's solo piano compositions. Dancer Bunny Briggs followed, accompanied by the band. Next was a "Memories of Kansas City" segment, featuring Hot Lips Page, and finally a "Suite of George Gershwin Ballads" was played by Bobby Hackett and Mel Powell. The concert closed with an "impromptu ensemble," a blues which evolved into W. C. Handy's composition "Ole Miss," on which everyone played. The closing blues became a hallmark of the Condon concerts which followed.

It seems hard to believe now that so many great jazz musicians could casually assemble in New York at these various events. Even in the midst of such a wealth of talent, Pee Wee's contributions stood out. At the end of the year, when the *Down Beat* readers' poll came out, Pee Wee had won the clarinet category—by thirty votes.

One unusual gig Pee Wee played around this time took place in the fur department of Gimbel's with pianist Dick Cary and Condon. Having them serenade the Christmas shoppers was undoubtedly an Anderson promotion. Another Anderson venture was scheduled to take flight in early 1943. *Orchestra World* in January reported:

> Coca-Cola is sponsoring an All-American All-Stars Overseas Jazz Band to tour overseas Army and Navy posts, possibly traveling by Army bomber. The outfit will carry on its jam sessions for the duration of the war. Tentative line-up for the all-stars at press time includes: Mel Powell, piano; Eddie Condon, guitar; Marty Marsala, trumpet; Pee Wee Russell, clarinet; Bud Freeman, tenor sax; Brad Gowans, trombone; Sid Weiss, bass and Ray McKinley, drums. In the event this line-up is carried through, it means that both Ray McKinley and Bud Freeman will be giving up their own bands. Another drummer being considered for the spot is George Wettling, now with Chico Marx.

The magazine added that staff writer Mel Powell would report on the all-star band "as it travels throughout the world in an exclusive series of dispatches." We do not have Pee Wee's reaction to flying all over the world from one military base to another for the tour never materialized.

Anderson had nothing to do with one major event in Pee Wee's life around this time. On March 11, 1943, Pee Wee and Mary were married at City Hall. The event was witnessed by Lillian Sheppard and Martin Katz, friends who lived in the Russells' apartment building. "We had an awful wedding," Mary told Whitney Balliett years later.

> Danny Alvin, the drummer, stood up for us. He and Pee Wee wept. I didn't, but *they* did. After the ceremony, Danny tried to borrow money from me. Pee Wee didn't buy me any flowers and a friend lent us the wedding ring. Pee Wee has never given me a wedding ring. The one I'm wearing a nephew gave me a year ago. Just to make it proper, he said. That's not the way a woman

> wants to get married. Pee Wee, we ought to do it all over again. I
> have a rage in me to be proper.

The newlyweds honeymooned at their West 21st Street apartment.

They had a special closeness that had developed out of their
bantering relationship. "She used to shout at Pee Wee a lot,"
said Marian McPartland. "He made faces suggesting silent tor-
ture." There was a lot to shout about. While Pee Wee had
developed a dependency on Mary, she was anything but a door
mat. Many of her social and political attitudes conformed to
prevailing attitudes in the Village. Although not politically active, she described herself as a "Red." (During that period, to be
sympathetic to Russia, America's wartime partner, was an article of faith among the Village's left-wing intellectuals.) She was
also a sometime believer in "free love," and, although she did not
like Pee Wee's occasional romps with female fans (a cartoon
celebrating his reputation with the ladies made a national men's
magazine around this time), she maintained a sophisticated attitude. She was a strong-willed, independent woman when it was
not considered "proper" to be one. She supported Pee Wee's
fragile ego and attempted to give him the personal confidence he
often lacked. She made a home for him.

Jazz concert fever spread to Philadelphia, where in June the
Basin Street Club of Philadelphia held its first event. Three
bands played from 1:30 to 6:00 p.m. The program was overly
ambitious, and the audience, although enthusiastic, was small.
"The club's angels had a little digging to do," reported *Down
Beat*.

Shortly after the concert, Pee Wee was involved in an automobile accident in which he suffered a serious leg injury. The
June 15, 1943, edition of *Down Beat* reported that the accident
occurred when an army jeep in which he, Chelsea Quealey and
Brad Gowans were riding overturned. Quealey and Gowans
were slightly injured. "At press time," *Down Beat* reported, "Pee

Wee had returned to work at Nick's, sporting a moustache and minus about fifteen pounds," a major weight loss for someone already so underweight. Another article in *Jazz Magazine* reported that Pee Wee had rejoined "Gowans' Nicksieland Band after a serious illness which necessitated a few weeks in the hospital." The army paid the hospital bill, which would indicate that the musicians were not responsible for the accident.

The band kept changing at Nick's. In July, Brad Gowans was the leader, with Tony Spargo, one of the members of the Original Dixieland Jazz Band, on drums. During the fall cornetists Wild Bill Davison, Bobby Hackett and Muggsy Spanier, trumpeters Marty Marsala and Sterling Bose, trombonists Gowans and Miff Mole, and pianists Dick Cary, Art Hodes and Gene Schoroeder kept coming and going while Pee Wee and Condon provided the continuity.

Mary Russell burst into print with a letter to the editor of *Time,* commenting on a story about the use of marijuana by jazz musicians. The piece was based on the incarceration of Gene Krupa in San Quentin for marijuana possession and said that it was a "violation which, if it could be universally detected, would land a great many jazz musicians behind prison bars." Nevertheless, the article was surprisingly favorable toward the use of marijuana, concluding that

> Lushes often die young from cirrhosis of liver or apoplexy, often spend their final days in delirium tremens. But vipers [marajuana smokers] frequently live on to enjoy old age. In "You Rascal You," a viper addresses an imaginary lush: "I'll be standing on the corner high when they bring your body by."

Mary was incensed at the portrayal of jazz muscians as either vipers or lushes:

> The author (I use the word advisedly) of the article "The Weed," a subtitle to your music (?) column, arbitrarily dumps jazz musicians into two categories. He gives us the hopheads, the drunk-

ards and, unfortunately for the future growth of American jazz music, no middle road . . .

While I have never observed in my husband the marijuana symptoms your staff writer so obligingly points out, I am well aware of the lamentable fact that all his friends are jazz musicians . . . I *have* seen my husband with more than one under his belt and alas, on occasion, one or two or even four of his friends harmonizing "Baby, Won't You Please Come Home" . . .

This is my problem. I'm the mother, sister, companion, nursemaid, sparring partner and wife of a jazz clarinet. Should I continue to live with this sink of iniquity and listen to his lousy rendition of "Baby, Won't You Please, etc." or should I divorce him, marry a shoe salesman or magazine writer, and listen to "Sweet Adeline?"

The editor responded:

Mrs. Russell should take a sober look at the facts. *Time* estimated that marijuana smokers among all U.S. dance musicians as not more than 20%, guessed that drinking dance musicians might outnumber the marijuana smokers. This still leaves plenty of room for sober clarinetists.

Sober or not, all musicians' recording activities had been curtailed for over a year as a result of the musicians' union's recording ban. When the ban finally ended in the waning months of 1943, Pee Wee was among the first musicians to record. The session was one of the first under the auspices of a young fan, Bob Thiele, for his new label, Signature Records. As the session on November 20, 1943, progressed, Pee Wee became more and more inebriated, finally requiring another clarinetist, Ray Eckstrand, to take over the solo on the last number, "Old Fashioned Love." That session produced little important music, but Thiele did not forget Pee Wee. He was still recording the clarinetist two decades later.

Much more important sessions were conducted in November and December, when Milt Gabler produced five sessions with a more steady Pee Wee. Two sessions were issued under Wild Bill Davison's name, two by Condon and one by Georg Brunis. All produced memorable results, but the performances of "Panama" and "That's A Plenty," by Wild Bill Davison's band, issued on a 12-inch disc, are masterpieces. John McDonough has written of Pee Wee's solo on "That's A Plenty":

> With the cunning of a cat chasing a mouse, he races through the first 16 bars in a frenzied double time. Without warning, he sweeps up to high F with a brilliant glissando, to resume the chase in the second chorus. Then, at mid-point, he abruptly changes course to float calmly above the beat, in half time, for the final eight bars. Even the most accomplished clarinetist would find this a difficult solo to play, and it would be impossible for anyone else to capture the inflections. Russell makes it sound easy.

Gabler said, "When I gave them a 'stand by!' and the red light went on in the studio, they were ready. They were off to one of the happiest days I have ever experienced." Two days later, the same front line of Davison, Brunis and Pee Wee recorded for Gabler again, this time producing one of the label's biggest hits, a Georg Brunis parody of the tune "Pretty Doll," which the gang had recorded with Fats Waller in November, 1940. With new lyrics, however, the old tune became "Ugly Child," and was a novelty hit of the year. A few days later, Gabler recorded Davison's group again, but this time it was for a new series of recordings that he had talked the World Transcription Service, a subsidiary of Decca Records, into making. The recordings were issued on 33 1/3 rpm, 16-inch discs and sold to radio stations for broadcast use only. In all, 43 transcriptions eventually would be issued as part of the "Jam Session" series, with many featuring Pee Wee.

The "Jam Session" series "actually became a time capsule of the jazz and small club scene in the United States for the year 1944," Gabler said.

All of the sessions were done separately from my Commodores, usually soon after the Commodore dates, so many of the personnels are similar, but they do vary due to the availability of musicians. The transcriptions had to be made in two-hour sessions, as the union regulations were different and my budget from Decca was a low one.

As the programs developed, I called a few special sessions for short length performances to round out the fifteen-minute shows and allow for commercials. Many of the dates were cut simultaneously for Decca and I issued some on their Brunswick and Decca labels. I also did many of the dates so that the men could earn extra money.

The series, with scripts written by Robert Paul Smith, were sold on a subscription basis to radio stations in May, 1945.

On December 18, 1943, another Town Hall concert was held, this time as a memorial for Fats Waller, who had died suddenly three days before. Art Hodes played some blues on the piano, Pee Wee was featured on "I Ain't Gonna Give Nobody None of My Jelly Roll," and Fats' mentor James P. Johnson played a selection of the late pianist's compositions.

Something old and new was being tried at Nick's under the direction of Brad Gowans: a re-creation of the Original Dixieland Jazz Band. At first the band included Wild Bill Davison, Pee Wee, Gowans and Condon. Condon left almost immediately, and soon Bobby Hackett replaced Davison. Shortly thereafter, Pee Wee left, and Gowans moved over to the clarinet chair, bringing in, on trombone, Eddie Edwards, one of the actual members of the ODJB. By the end of October, one provocation or another inspired Nick to fire the entire band, mainly to get rid of Gowans. Gowans immediately signed a six-week contract to play a featured spot in Katherine Dunham's "Tropical Review," set for a nationwide tour.

The year 1943 ended with Pee Wee again winning the *Down*

Beat readers' poll in the clarinet category, this time by 1,632 votes. The Commodore recordings, the concerts and the publicity Pee Wee received in the jazz press all contributed to his popularity with the fans.

One of Pee Wee's most fervent fans at Nick's was the foreign correspondent Eddie Gilmore, who was assigned to Moscow by the Associated Press when the war broke out. Gilmore was an object of one of Pee Wee and Mary's hobbies—making long-distance telephone calls. Friends and fans all over the country would occasionally get a call from them for no apparent reason. This particular call, however, was quoted by Broadway columnist Earl Wilson—translated into the peculiar entertainment column lingo of the time: "We hadda make a long distance call," Wilson reported Pee Wee told him. "Y'ever get like that? When you got to make a LONG distance call?" As Mary told the story: "Pee Wee's tired. He's so great, but he's lazy. Well, that night— that morning—he says, 'Can I call somebody in California?' I say, 'You got a friend in Russia; what's the matter with calling Moscow?'" They assembled several other musicians and called Gilmore. The startled and overjoyed newspaperman heard a spirited version of "I Ain't Gonna Give Nobody None of My Jelly Roll." The call cost $68. Pee Wee and Mary felt it was money well spent, a contribution to the war effort.

At the season's second concert at Town Hall on January 8, 1944, Pee Wee was presented with his *Down Beat* trophy by singer Carol Bruce. Pee Wee was still a regular in the band at Nick's, now under the nominal leadership of Miff Mole. The pace of Pee Wee's outside work was increasing rapidly. On February 9, he recorded with Mole's band for the World Transcription series. On February 19, the third Condon concert of the season was held at Town Hall, featuring the "new band at Nick's." Pee Wee was featured on "Squeeze Me."

On March 4, Pee Wee recorded with a quartet under the leadership of Nick's intermission pianist, Cliff Jackson. The records were poorly recorded for a new label, "Black and

White," and they received poor reviews at the time. The February, 1945, issue of *Metronome* reported that "one of the musicians who made this session told us: 'Man, that was the saddest date I ever played in my life.' We can sympathize with him. Pee Wee Russell is at his least good, shall we say, and Cliff Jackson is by no means at his best." Heard today, however, Pee Wee's playing on this record sounds exciting and innovative.

On March 8, the Condon gang recorded a session for the army's label, V-Disc. The records were shipped to military outposts all over the world to be played as a morale booster for servicemen. On March 11, the concert at Town Hall was enthusiastically reviewed in *Metronome,* a magazine usually hostile to Nicksieland.

> While the previous one was the least of the Eddie Condon Town Hall jazz balls, in this reviewer's opinion, this was one of the best, along with the first in this season's series. The opening group, "Songs Played at Nick's," spotted a rough ensemble (Maxie Kaminsky, Miff Mole, Pee Wee Russell, Cliff Jackson, Pops Foster, George Wettling and Eddie Condon) in "Darktown Strutters' Ball," "Dear Old Southland," "Jada" and "Muskrat Ramble," Maxie sounded fine above the ragged group in all four numbers. Bobby Hackett came in for "Jada," played at a fetchingly slow tempo, with some lovely middle register Hackett horn and better, more evenly pitched Russell clarinet than usual.

The concert was recorded by CBS and a portion of it broadcast over WHN as a trial run. The next day the group recorded for V-Disc again, but this session remained unissued for more than forty years, when it finally appeared on LP.

An excursion to Washington, D.C., almost proved to be a disaster. The band had been booked into Constitution Hall, then operated by the Daughters of the American Revolution. But when the patriotic ladies discovered that the group included some black performers, the contract was cancelled. Condon went public with a press conference. As he recalled in his memoir, "At the last moment we were informed by the Daughters of the Awful Right that we could not play in their hall because we used 'mixed' bands and might draw 'undesirable' elements." They performed instead in the ballroom of the Willard Hotel on March 26 to an overflow crowd.

Back in New York, more recording awaited. World Transcription sessions were made under Condon's name on March 30 and under Muggsy Spanier's leadership on April 4. On April 8, another Town Hall concert was held.

Gabler recorded the gang for his Commodore label on April 15, April 22 and April 28 under the name of Muggsy Spanier for the first two sessions and Miff Mole for the last. The Mole band, with Hackett, was holding forth at Nick's. Pee Wee's performances on all three sessions are consistently electric, with exciting ideas bursting from his clarinet in fascinating contours and shapes. But on Mole's "Peg O' My Heart," he gave one of the most soulful performances. The following day, April 29, the gang convened another clambake at Town Hall. The Town Hall affairs even received notice in *Scholastic,* a magazine for high school students. In an article entitled "Jazz Lover's Paradise" by Herman L. Masin, the students were told

> Right now, the greatest mecca for jazz is Town Hall in New York City . . . only five people could assure sell-out crowds. They were: Marian Anderson, Arturo Toscanini, Lotte Lehmann, Vladimir Horowitz and Dorothy Maynor. This year, a sixth was added—Eddie Condon! . . . Eddie dislikes putting labels on jazz. To his way of thinking there are just two kinds of jazz— good and bad.

Predictably, *Metronome* was not upbeat.

> Eddie Condon's last two jazz concerts of the season were somewhat disappointing. A paucity of talent characterized the affairs . . . The opening number of the April 29 concert was entitled "An Evening on Fifty-second Street 1935," which featured Miff Mole, Max Kaminsky, Pee Wee Russell, Frankie

Froeba, Eddie Condon and Joe Grauso. "Sweet Georgia Brown," "The World Is Waiting for the Sunrise," "Cherry" and "That's A Plenty" were their numbers, played roughly, with some fair solo moments." . . . Ed Hall, Rex Stewart, Pee Wee Russell, Ernie Caceres offered viable solos, with backing by Joe Bushkin, Oscar Pettiford and Sonny Greer . . . On the 13th of May, the sun shone brightly, mightily and scared away a lot of the fine musicians scheduled for the Condon concert. Result was a bedraggled Nicksieland opener.

Despite *Metronome,* the concerts were successful enough with the fans to induce the Blue Network (ABC) to schedule them on a sustaining basis (although Chesterfield Cigarettes was a possibility for a time, no sponsor was found), and the Armed Forces Radio Service began to transcribe them for replay on their overseas network. The concerts were broadcast live on Saturday afternoons beginning May 21, 1944, and continuing until April 7, 1945. It had not been easy getting the network interested in carrying the concerts. Recalled Anderson:

> My proposal for an Eddie Condon Jazz Concert series had been turned down by all the networks including the Blue. They would assess that the advertisers would object to: 1, the origins of the music, the red-light district of New Orleans; and 2, they were also firmly against any mixing of the races. Since there was no prospect of getting any paid advertising to sponsor such a show, they had to reject the attraction as impractical. . . . Then one day, Esmé O'Brien, a jazz fan and then-wife of Bobby Sarnoff, son of the president of NBC, heard about it and she told Bobby. Suddenly, NBC's Blue Network reversed its previous decision and scheduled us for the series. That's how we got on. [Esmé O'Brien's second husband was the noted jazz producer John Hammond.] . . . For the appearances at Town Hall, each musician was entitled by the union to $18. I paid everybody $25, usually in cash in a Town Hall ticket envelope.

The enthusiasm of the fans in the studio was communicated to the listeners. The re-broadcasts by AFRS to the servicemen were a sensation. The programs were voted the most popular in both the European and Pacific theatres, even topping the Hit Parade program in Europe. Pee Wee was heavily featured on the shows, and Condon frequently made comic references to the clarinetist, calling him "the thin man of jazz," "the commando" or "the pin-up boy," among many other sobriquets. Pin-up photos of Pee Wee were offered free to servicemen who wrote in—and many did.

"We tried everything to get that mail coming in," Anderson recalled. "We wanted to try to get a sponsor for the shows." Although Anderson was not successful in finding a sponsor, the shows spread Pee Wee's fame "from Ketchikan to Calcutta," as the announcer put it, and not just with American soldiers. A group of R.A.F. flyers wrote to Pee Wee that they always played his records "for courage" before taking off on a mission. Condon may have been the witty, fast-talking front man, but it was Pee Wee many of the fans came to hear, and Anderson knew how to capitalize on that too. Eddie Condon's name was on the posters, but the photograph was of Pee Wee.

The Town Hall concerts provided jazz with one of its periodic brushes with popular acceptance just when the jazz world was being rent asunder by a new movement called "bebop." The animosity on both sides was inflamed in large part by critics, especially those at *Metronome,* who championed the new music primarily by attacking the old. The popularity of the Town Hall concerts made them an easy target. Many contemporary swing stars, including Tommy and Jimmy Dorsey, Gene Krupa, and Woody Herman, appeared on the Town Hall shows, and the concerts ultimately were derided by the more "progressive" critics for being old fashioned and repetitious. The traditionalists, however, had their own press and fought back.

Reviewing the first concert, Art Hodes in *Jazz Record* captured the breathless astonishment of the Nicksielanders that greeted the broadcasts.

Eddie Condon—NBC—3:30 on Saturday—a national hookup—
a jazz program. Fred Robbins introduced Condon, who in turn
presented Max Kaminsky, Pee Wee Russell, Miff Mole, Joe
Grauso, Gene Schoreder, and Casey in a band number. Then
James P. Johnson, piano solo—John O'Hara and Condon reading
script, Lips Page singing and playing—some gal vocalist (this was
the low point of the program) and a bit of a jam session, lightly.
The announcer said it would run for 13 weeks.

The band at Nick's was drawing bigger crowds than ever, in
part because of the mushrooming popularity of the concerts and
the broadcasts. Mole was the nominal leader, with Sterling Bose
on trumpet. (Bobby Hackett replaced Bose now and then during
this period.) The Tuesday night jams drew the biggest names in
jazz from big bands and combos whenever they played New
York. Pee Wee's old boss, Red Nichols, then featured with the
Casa Loma band at the Hotel Pennsylvania, sat in one night and
"played the can off Bobby Hackett and Sterling Bose," according
to one eye-witness Nichols partisan.

Pee Wee left Nick's in July 1944 for a vacation, but he and
Mary could never quite get organized enough to get out of town.
They had started corresponding with Jerry and Wanda Simpson,
two fans from Saginaw, Michigan, who had invited them to
vacation there. "I'm on a hell of a vacation," Pee Wee wrote.
"I'm sorry I'm not with you but I had to stay in town—
broadcasting. Mary and I are having a bitch of a time: zoo,
feeding pigeons, Chinatown, movies, drinking, romancing. I've
forgotten what it felt like but it's nice. Good night now. Uncle
Pee Wee."

In August Pee Wee teamed up with James P. Johnson's band
at the Pied Piper. He returned to Nick's in September; the band
included Muggsy Spanier and Miff Mole. There were more
sessions for Commodore in September and one for Associated
Transcriptions, another radio station service. Pee Wee's celebrity status led another instrument company, the Woodwind

Company—apparently unaware of the Conn clarinet fiasco—to seek his endorsement. Ads appeared in *Down Beat* in September: "The mouthpiece used and recommended by Pee Wee Russell is the original Woodwind model."

At the September 23, 1944, Town Hall concert, announcer Fred Robbins mentioned a theatrical project that was in the works. Composer Vernon Duke and lyricist Howard Dietz were putting the finishing touches on their score for a new Broadway musical based on the Somerset Maugham play *Rain,* renamed *Sadie Thompson.* The plot required the title character to play a jazz record from time to time, and Duke and Dietz had written a new tune for the recording—"Poor as a Churchmouse." According to Robbins, Duke approached the boys at Nick's to record the tune for use in the show. In all probability, the tune was played by the pit band, but if the recording was made, it has not been found. The gang did play the tune at the concert.

Gabler featured Pee Wee on another Commodore session on September 30, accompanied by Jess Stacy, Sid Weiss and George Wettling. It was the first session to be issued under his name since the HRS session in 1938, and produced several memorable discs. Stacy said:

> Working with Pee Wee was fun. It was easy. With Goodman, you'd rehearse and rehearse and rehearse until your fingers were so damned beat you'd think they were going to fall off. With Pee Wee, we'd decide what we wanted to do, then we'd go through the number a couple of times, and then we just played it. That's why the records had that fresh quality. We were equal partners in an ensemble. I could get behind Pee Wee and really push. With Benny, I always thought if I forced anything, I might be irritating him. With Pee Wee, I could get in and counterpunch and play creatively.

The session was not without incident, however. "Mary was at the session," Stacy recalled.

Pee Wee had a jug of booze with him. Well, we were doing a rehearsal take on "Take Me to the Land of Jazz," I think. And while we were doing it, Mary snuck around and stole his jug and left. As soon as Pee Wee noticed it was missing—and that was pretty quick—we had to stop the session, take a break while somebody went out and bought Pee Wee another bottle. It was 45 minutes before we got back to work. Pee Wee was seething. Mary wasn't trying to get him to quit drinking—she wanted it for herself. Milt thought it was pretty funny.

At the beginning of October, the Russells found a new apartment. "We've moved!" wrote Mary to the Simpsons:

But only after Rod Cless and Sterling Bose descended on us one morning and moved us bodily. At any rate, it was practically that. And we're having ourselves a time. The Russell walks around anxiously emptying ash trays, if you can picture such a thing, and carefully picking up stray matches and bits of fluff from the floor. He's interested in lamps and shopped in Hearns today for garbage cans and mops and such things. Very nice. Any day now you may expect to find two decent, law abiding citizens at 205 West 10th Street.

There's been stuff goin'. So much in fact, that I've divorced the whole business as unbecoming to the wife of a clarinet player. A real, honest to goodness feud divided into bitter camps. Pro-Pec Wee and anti-Pee Wee. Muggsy and Miff are the two foremost gents solidly behind Pee Wee. So much so that they offered to fluff off Town Hall which incidentally has moved its weekly broadcast to the Ritz Theatre.

Here is the whole sad story. It began with a secret record date. Secret for the same reason that Fats called himself Maurice on several records. Gene Krupa has contract commitments and can't record for any company but his own except under another name. Somebody had a brainstorm and decided that nothing could be better than trio records with Gene, Jess Stacy and Pee Wee. Which is a pretty good idea for everybody but the Peach who can't get away from the office at midnight to record and Ernie Anderson who more or less runs things, seemed to think that was damned ungrateful of him. What the hell! So what if Nick fires you, old man, what the hell. He'll take you back in three or four days, won't he? Okey, Ernie darling, and what are you paying Gene Krupa. But he's Gene Krupa!!! And what are you paying Jess Stacy??? Oh come now Pee Wee, he's Jess Stacy. Well, goddammit, I'm Pee Wee Russell. Oh, it's a swelled head you have now. After all we've done for you. Sure, I'll use Ed Hall on everything. I'll see that you never get another job in New York. I didn't want you in the first place. After all we've done for you.

So that's it. They've done a lot for him. He's made records. Sure. For scale. Joe Jones from the Bronx makes the same thirty bucks for a record date. Yeah, they've done a lot for him. Altruists, that's what. They hire him out of the goodness of their hearts. His clarinet has nothing to do with it. The letters buying records because he's on them has nothing to do with it. They use him because he's pretty, because they want to give him a break, because he's a good kid. And he'll never have another record date as long as he lives.

So the first Saturday he didn't have the broadcast [Ed Hall was the clarinetist on the September 30 broadcast] he worked all that afternoon making records as the nominal leader, with Jess Stacy, Sid Weiss and George Wettling. He made Land of Jazz (and I love that especially because I dug it up), Rose of Washington Square, the D.A. Blues and something I can't remember. And I'm doing a small burn. I've never been madder in my life. Not because of his business troubles, not because he developed a nervous stomach and didn't eat and didn't sleep because of his deep sorrow (he thought they really liked HIM) but because people people and sweet people (I mean that) like Maxie Kaminsky and other nice guys arbitrarily decided I was doing it. That Pee Wee wouldn't stick up for himself on his own. Only they didn't call it that. It was called a swelled head. And that I was prodding him. They didn't recognize the pride. They've been too used to shoving him around. Oh, in a nice way. And it won't be necessary to tell you, Wanda, that I've never interfered in Pee Wee's work in any way. I don't feel I owe anybody an explanation. Nor does Charles. So we say nothing. And that's the way it stands.

It's really a divided camp and stuff going and people fighting. Especially the Carnegie Hall deal. There's a concert at Carnegie Hall next Monday night. A real stuffed shirt stuff. At three bucks a throw and the best thirty jazz 'artistes' in the world. It's all very pretty and very la de da. The posters have plastered the city. There's one guy's picture on those kids. It's Pee Wee. And nobody asked him. Nobody invited him.

Apparently, feelings were smoothed, because Pee Wee *was* featured at the Carnegie Hall concert, October 14, 1944, the first of the season. *Down Beat* announced that the concert would include "30 jazz stars" and would feature Gershwin's "Concerto in F," arranged by bassist Bobby Haggart and featuring him, Pee Wee, and Billy Butterfield, who played the part that Bix Beiderbecke had immortalized on the Paul Whiteman Orchestra's recording sixteen years before.

The reviews of the Carnegie concert were mixed. "Deeply Disappointing" was the headline for Barry Ulanov's report.

Miff Mole contributed his familiarly facile, cleanly lined "Peg O' My Heart" (wherein Pee Wee played some pretty clarinet) . . . Personal taste, perhaps, governs my lack of enthusiasm for the work of Pee Wee, or Art Hodes and Willie the Lion Smith and Red McKenzie, Joe Grauso and Bob Casey and Bob Haggart, at this concert . . .

The reviewer for *Jazzette* was only slightly more upbeat:

The whole concert was a bit disappointing and definitely handicapped by the immensity of the hall. It was impossible to capture the spirit and intimacy of Nick's, the Pied Piper and Jimmy Ryans. Yet the overall quality of the music and the opportunity offered the public to hear the "greats" of jazz more than made up for the other minor deficiencies.

Ralph Berton's review in *Jazz Record* pointed to one of the problems with the concert:

. . . with few exceptions most of the numbers were too short.
Often only a few choruses were played, practically never more
than two consecutively by the same man . . . Pee Wee Russell
was good, as usual, though seldom really inspired. No one
seemed to relax . . . "Peg O' My Heart" wasted both Pee Wee
and Muggsy, cluttered up with a strictly George Olsen riff which
was a real bringdown . . .

But the review was basically positive, concluding, "all in all,
your money's worth and then some."

Leonard Feather was also encouraging, if patronizing, in his
review the same month in *Metronome* about the Condon broadcasts:

Eddie Condon's jam sessions have gone into their second
thirteen-week stretch on the Blue, and their musical standards
have gone up since I covered this show shortly after its inaugura-
tion. The main improvement has been in the drum department . . . when the front line includes, say, Kaminsky, Miff
Mole, Pee Wee or Muggsy, Miff Mole and Ed Hall, the jammed
ensembles are well integrated and everybody plays the right
chords . . .

George Simon chimed in at *Metronome* about a subsequent Carnegie concert:

"I like Dixieland. I like Nicksieland," he protested. In fact, I
guess I'm one of the few critics of modern jazz who'll fight to the
last ditch for good dixieland and good Nicksieland. Maybe that's
why I'm so distraught about the Eddie Condon concert. Right
from the start, the smell of clambake was imminent.

The review went downhill from there.

Simon's next piece in *Metronome* painted a picture of the
scene at Nick's at the time. The clientele still included a pan-
orama of record collectors and Greenwich Village artists, writers
and actors, but now the scene included servicemen:

"He sounds like he's got a split lip," said the man next to me at
Nick's, while a young soldier standing directly in front of Pee

Wee went into ecstasies as Russell finished his clarinet passage on "Muskrat Ramble." It struck me as being pretty indicative of the reactions of the majority to Dixieland: either you think it's the greatest or you're convinced it's terrible . . . There are too many young squirts who have been attracted to Dixieland not because of the music but because of the glamour they've read about. I'll bet you anything that Pee Wee's face is far more important to them than what he plays, and, commercially, I dare say, it's probably far more important to Pee Wee too! . . . Dixieland no longer has to be great to survive. No, Pee Wee just has to make a few faces, Condon just has to make a few nasty cracks about big bands, Muggsy just has to look sad, and nobody has to keep time anymore if he doesn't feel like it.

That was too much for Mary. After a skirmish with the irate wife, Simon diplomatically retreated in his next column:

"What'd you have to write that article," asked Mary Russell, who's Pee Wee's wife and an awfully nice person. "It's not nice to criticize a man's playing because of his face, and besides, Pee Wee's been taking an awful lot of kidding about it." I pointed out, with some degree of conviction, I hope, that I have and had written nothing against (1) Pee Wee's playing . . . (2) Pee Wee's face, which has a great deal of character, but which, I pointed out in the article, has been over publicized, so that consequently too many ickies don't honestly appreciate Pee Wee's clarineting but instead just come to Nick's to see, instead of to hear, him. "I can't help it if I make faces when I play," added Pee Wee with a touch of sadness. "I know I'm not the best looking man in the world, but that shouldn't be held against my playing." I honestly feel sorry if I have subjected Pee Wee to any embarrassment.

Making further amends, Simon featured Pee Wee on a V-Disc session he produced under the leadership of Muggsy Splanier. The session on October 17 included Pee Wee's first recorded vocal on a blues appropriately named "Pee Wee Speaks." Another Associated Transcription session under Condon's leadership followed on October 24, and he closed out the month with a "For the Record" broadcast with Miff Mole on October 30. The radio program was designed to promote the artists who appeared on V-Discs (all without compensation) and to generate material to be issued by the armed services. On November 3, Pee Wee received a letter from Captain Robert Vincent of the Army Special Services Division.

> I would like to thank you once more, on behalf of all of us here at "V" Discs, as well as the millions of men overseas whom we serve, for your fine contribution to our "For the Record" show on Monday. It was indeed a fine gesture on your part and one, which I am sure, will make many of our fighting men very, very happy.

Pee Wee always treasured the letter, feeling that he had at least done something for the war effort.

On November 20, Pee Wee gamely climbed into an airplane and flew to Montreal with Art Hodes, who had already conducted a successful jam session concert there at His Majesty's Theatre on October 2 with Mezz Mezzrow on clarinet. "Can you imagine what it was like being in a plane for 45 minutes with Pee Wee while we circled over the airfield because the weather was too rough to land?" asked Hodes plaintively in his next column in *Jazz Record*. The flight was bumpy and nerve-wracking for the musicians. "The plane came close to crashing," Mary Russell reported. "Danny Alvin told me he said his prayers and meant it for the first time in his life."

Perhaps because of the flight, Pee Wee and the other musicians were not in the best condition for the concert, according to a review by Roy Ervin the following day in the *Montreal Star*. "Pee Wee Russell, clarinet, who was expected to be the evening's star," wrote Ervin, "spent most of it posed like a drooping Pan, hid quietly behind the piano's open top or disappeared from the stage altogether. These sessions, you see, are QUITE informal."

From Montreal, Pee Wee was to have traveled to Detroit to perform there with Muggsy Spanier's band, but the gig was cancelled at the last minute and Pee Wee returned to New York in time to appear at the next Carnegie Hall concert.

The informality of the concert format noted in the review of the Montreal affair was irritating to the increasingly critical jazz press as the battle lines were drawn between the boppers and the "moldye fygges." Reporting on the December 2 Carnegie Hall concert, Barry Ulanov wrote:

> At seven o'clock the scheduled intermission was indefinitely postponed, because circumstances at the concert were beyond Eddie's control, and I left. An hour and a half of thin ensembles, solos I had heard many, many times before, and tunes that were all too familiar, was enough of too much . . . Informality is fine to a degree; carried to the point of collapse it is pointless. When the intermission is omitted, numbers run way over schedule and there is an almost unceasing monotony of sound, I think a concert has collapsed.

Leonard Feather was only slightly more charitable about the Condon broadcasts from Town Hall and the Ritz Theater. Noting that the radio broadcasts "were the first of this kind to go from coast to coast," Feather admitted they had "done much for jazz as a whole:

> It could have done a great deal more had Condon not been tied rigidly to the Dixieland instrumentation and insisted on a policy of playing Dixieland tunes, often against the wishes of the musicians themselves. Thus no alto or tenor saxmen, no guitarists and relatively few colored musicians have benefited from the substantial network time allotted to these shows.
>
> Many of the musicians themselves feel that opportunities are being wasted at these Condon affairs, which in addition to the broadcasts from Town Hall and the Ritz Theatre, included four unbroadcast concerts held at Carnegie Hall. After the first of the Carnegie affairs, I talked to a number of musicians. The general opinion was that the affair had been a disorganized clambake. Edmond Hall confessed that he never feels relaxed, can never play at his best, at such sessions; Joe Marsala admitted he hadn't

wanted to do "Wolverine Blues," but they had insisted he do this Dixieland tune. Hot Lips Page, Sammy Price, Muggsy Spanier, Benny Morton, Sid Weiss and several others told me they had been uneasy and unable to play their best under the circumstances. What this all boils down to is that there are a few musicians who are born Dixielanders, but many more have Dixieland thrust upon them. Such men as Spanier and Wettling have often stated that they dislike being classed as Dixieland musicians. But the individual ability of some of these men sometimes saves the concerts and broadcasts from being a collective mess.

Despite the negative press, the concerts and broadcasts were more popular than ever. Several of them in December were especially memorable for Pee Wee's reunion with Jack Teagarden and Wingy Manone. The musicians appeared on other radio programs as well in order to plug the concerts. One memorable broadcast was on the Reader's Digest program when Condon attempted to explain what jazz was, illustrating his points with a blues played by the band which included Pee Wee, Bobby Hackett and Jack Teagarden.

Anderson was branching out, presenting the concerts in other cities.

We played Symphony Hall in Boston and the Academy of Music in Philadelphia many times. We also often took our players to Princeton University and made many appearances at Hamilton College in Clinton, New York. Princeton had a long-standing committment to good jazz. Only a few years earlier they had brought the whole Fletcher Henderson band to the university for a concert. This was bold, for the band was black and at that time blacks just weren't seen at Princeton. The racial problem was solved by having the whole orchestra screened by potted palms. In our day I can't remember ever playing there without some blacks in our group and all had the run of the place. Invariably the students entertained us lavishly.

Pee Wee finished the year with another burst of recording for Gabler, on December 7 under Muggsy's name for Commodore, on December 8 with Red McKenzie and Jack Teagarden for World Transcriptions, with Condon on December 13 for a Decca session supervised by Gabler, and on December 14 with Teagarden on a Condon session which was issued both by Decca and World. Amid all the musical activities, on December 8, 1944, a tragic accident took one of Pee Wee's compatriots, clarinetist Rod Cless. Mary wrote to Jerry and Wanda Simpson:

> He was found on the ground floor about eight o'clock on Monday morning where he had apparently been lying since about four. It was assumed that he had taken a header from the second or third floor over the banister when he came in from work. And that, my children, is the evil of drink. I haven't touched a drop since October and I don't intend to.
>
> At any rate, I gave the hospital a pint of blood on Wednesday. His wife (Renee) brought Pee Wee his clarinet this afternoon. No histrionics connected with it. She was simply practical. She said Rod didn't like that kind of waste and she thought Pee Wee ought to have it. All the guys came through. They raised a couple of hundred dollars immediately so he could be moved to a private room but he died before they got around to doing it . . . Rod's wife is going to sleep on the couch here for a couple of nights. Being alone isn't good for her.

Despite having no manager or personal press agent (Ernest Anderson probably came closest to filling those roles, but on an informal basis), Pee Wee had succeeded in gaining an enormous amount of publicity and in attracting a large, knowledgeable and enthusiastic body of fans. He even figures in a *New Yorker* short story by S. J. Perelman, "Farewell, My Lovely Appetizer," appearing in the December 16 edition. It was a send-up of a Dashiell Hammett type of detective story in which one of the characters snarls: "You got the wrong pitch, copper. That stuff is hotter than Pee Wee Russell's clarinet."

Walt Disney also became interested in Pee Wee. "He came to several of the concerts," Ernest Anderson recalled,

> and then brought his brother, Roy, who ran the financial side. After one of the concerts, he told us he wanted to do a series with a character built around Pee Wee. The cartoons would use Pee Wee's voice and clarinet. We had several meetings on that in Roy's office in Radio City. Pee Wee said he would do whatever we said. Roy wanted to close the deal, but it was always postponed.

Eventually, the project was permanently shelved.

The Russells celebrated the wartime Christmas with an exchange of presents. "He gave me a blue silk or near silk nightgown," wrote Mary to Jerry and Wanda Simpson, "five pairs of stockings and a fur coat. I love my coat. I gave him a thoroughbred cocker spaniel, name of Weepy Muggsy McSpaniel. Very sad dog. Muggsy has been trying to win him away from Pee Wee. I think Pee Wee and Muggsy love the dog wildly. Crazy, cute bad tempered little thing. He likes me."

With Pee Wee's continuing success on the broadcasts and with the large number of outstanding recordings he made during the year, he had no trouble winning the *Down Beat* clarinet category again. It had been a good year.

Sixteen-year-old Pee Wee with his Conn saxophone. (Photo courtesy of Michael Goodman)

Herbert Berger's Orchestra at the St. Louis Club (Pee Wee is sixth from left). (Photo courtesy of Duncan P. Schiedt)

Pee Wee, Mezz Mezzrow, Sonny Lee, Bix Beiderbecke, unknown, and Eddie Condon. (Photo courtesy of Duncan P. Schiedt)

Louis Prima and his "Famous Door Orchestra" goes Hollywood in a still from a lost short.

Pee Wee solos with Bobby Hackett's short-lived big band. (Photo courtesy of Wayne Knight)

An unidentified angel of mercy arrives none too soon for Pee Wee and Fats Waller.

Pee Wee's unique "sound of surprise" breaks up Eddie Condon and Bobby Hackett at a Milt Gabler jam session. (Photo by Charles Peterson, courtesy of Don Peterson)

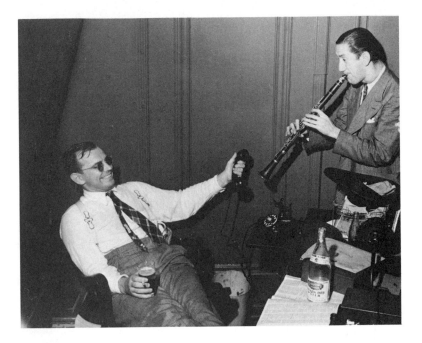

Ernest Anderson produces a phone for Pee Wee to serenade a long-distance fan. (Photo by Charles Peterson, courtesy of Don Peterson)

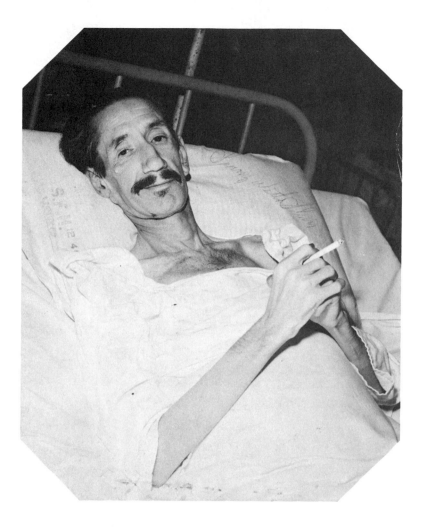

In the charity ward of San Francisco Hospital, January 24, 1951.

Pee Wee's first handle-bar moustache and his first band with John Dengler and Eph Resnick. (Photo courtesy of John Dengler)

Pee Wee and Mary at Newport. (Photo courtesy of Vera Goodman)

Pee Wee's art decorated the walls of his apartment for a few years.
(Photo courtesy of Michael Goodman)

THE PEE WEE RUSSELL QUARTET WITH MARSHALL BROWN

Charting a "new groove" with Marshall Brown

Instant rapport: Gerry Mulligan and Pee Wee at the Monterey Jazz Festival. (Photo courtesy of Ray Avery)

Joe Sullivan, Charlie Teagarden, Pee Wee, and Jack Teagarden,
reunited for the last time at the Monterey Jazz Festival. (Photo courtesy
of Ray Avery)

Enjoying semi-retirement: a few days in New Jersey. (Photo courtesy of Michael Goodman)

9 • • • •

__In Decline

*I*t was inevitable that the novelty of the big jam sessions would wear off and that the music itself would begin to pale. It may have seemed like a dream to be able to hear Pee Wee Russell, Hot Lips Page, Muggsy Spanier, Miff Mole, Brad Gowans, James P. Johnson and so many others at Gabler's sessions in 1938, but by the beginning of 1945 they had become all too familiar. The war was ending, and the public was beginning to consider the future and the allure of the new. Although the signs of decline were becoming evident in 1945, the popularity of Nicksieland was still strong. Condon continued the series of jazz concerts at the Ritz Theatre, broadcast coast-to-coast and transcribed for rebroadcast around the world. Local "barefoot gangs," as Condon named them, had been established in many parts of the country, a tradition that today has spread throughout the globe. The movement still had some momentum, but friction with the boppers and the jazz press was grinding it down.

The Carnegie Hall concerts continued to be sellouts. "Eddie Condon's Carnegie Hall concert of January 20 probably will present several of the musicians who have won top spots in *Down Beat*'s All-American band poll," the magazine reported, "though just who they'll be depends on who shows up—such is the studied informality of the Condon concerts. Pee Wee Russell is a virtual certainty but, as for the rest, you pays your money and takes your chance."

During the concert, Lee Wiley presented Pee Wee with his *Down Beat* trophy for winning the 1944 Reader's Poll for clarinet for the third year in a row. As the singer handed the award to Pee Wee, he was not impressed. "Let's see," he said on the broadcast, "I got two dollars for the last one and two dollars for the one before that. How much do you figure I'll get for this one?" On closer examination, the clarinetist noticed the trophy was inscribed to saxophonist Toots Mondello. It was May before Pee Wee finally received the correct award. "I persuaded Pee Wee to go to the *Down Beat* office for his trophy," Mary wrote to Jerry and Wanda Simpson. "And, believe me, it took a lot of persuading. When they gave it to him his name (as I pointed out to him) was spelled wrong and he assured me that they'll do better with the spelling next year and immediately turned scarlet when he realized what he had said. But everybody loved it."

In addition to all of the concerts, jam sessions, private parties—including several trips to Washington, D.C., to perform at jazz parties hosted by Ahmet Ertegun (the founder of Atlantic Records)—broadcasts, and recordings, Pee Wee continued at Nick's with Miff Mole's band, which, in February, 1945, included Muggsy Spanier. *Down Beat* noted the end of the season's Carnegie Hall concerts. The radio broadcasts from the Ritz Theatre, however, were renewed for another thirteen weeks. Meanwhile, Condon was planning a tour of the East Coast, with the first concert to be held in Boston's Symphony Hall.

Pee Wee still appeared to be having money problems with Condon and Anderson. "Pee Wee is through (again) with the Condon, Anderson combination," Mary wrote to Jerry and Wanda Simpson.

He's mad. He wants to get paid and nobody ever pays him. Of course the Blue [network] pays off every week but Anderson gets the check and neglects to pay off. And he's only paid Pee Wee for two Carnegie things. Pee Wee never did get paid for the Town Hall deals. Oh well, that's the way it goes. The boys are mad again . . . When Pee Wee finished his Blue broadcast instead of going on to rehearsal for Carnegie he went home vowing he was through forever. You won't be hearing him anymore on Saturday afternoons.

But, according to Anderson, not only was Pee Wee paid for his appearances, but Anderson frequently slipped him an extra five or ten dollars. Pee Wee was known to be very irresponsible with money and frequently was in a confused mental state. And it is also possible that, after being paid and spending it, he simply told Mary that they were holding out on him. It is impossible to discover the truth at this point, but in any case, Pee Wee seemed never to have any money.

Nick Rongetti decided to cash in on the continuing popularity of his house band by starting his own label, Manhattan Records, and recording them. Three sessions, with slightly varying personnel, were made, all on March 1 and 2, producing 18 sides. The records were released in three albums, each containing three recordings. All bore the legend "Nick Presents His Dixieland Band under the direction of," then naming Miff Mole, Muggsy Spanier or Pee Wee Russell. They were available for sale only from the hat-check room at the club. Because the wartime shortage of shellac was still acute in the United States, the records were manufactured in Canada. The sessions were fun for the musicians and for Rongetti, who even sat in on piano on "That's A Plenty," the tune he frequently played on one of the three pianos at his club. A blues in Muggsy's album was titled "Feather Brain Lament," in honor of one of their severest critics. "Did I tell you that Muggsy beat up Leonard Feather a couple of weeks ago," wrote Mary to Jerry and Wanda Simpson. "And that stinky Russell butted in too. I think he spat on

him (before Muggsy socked him) or something equally disgust-
ing. Somebody wrote a poem: "Two Face on the Barroom
Floor."

> Eddie Condon went and hired a hall.
> He figured all would have a ball.
> But a British boy, slow on the drawl,
> Wound up less wind-up in a brawl.
>
> Seems he's a critic name of Lennie
> Who's strictly jump and jive and Benny;
> And his review wasn't worth a penny,
> It was only one out of many.
>
> As fate would have, a few nights later
> At Pesky's place this two-time traitor,
> Met Muggs and Pee Wee (no two greater),
> Who were set to get this Dixie hater.
>
> So pretty soon began a tussle
> The first to poke at him was Russell
> Though long on words, Len's short on muscle
> Soon set by Spanier on his bustle.
>
> Now, who was it that really sinned?
> The boys who got their knuckles skinned,
> Or the lad who had his bad ears pinned
> As a white Feather flew in the wind!

Pee Wee's Manhattan album included one of his compositions, "Mama's in the Groove," which he wrote with trumpet
player Al Bandini. Later, Bandini added words to the tune and
called it "I Need You." Bandini was a regular at Nick's—his
brother-in-law, Andy Gardella, had been a bartender there—
and Al was a big fan of Pee Wee's. For several years when
Bandini led a dixieland band that played on a float in Macy's
Thanksgiving Day parade, he always made sure the band included Pee Wee. In addition to playing Armstrong-inspired
trumpet, Bandini also played drums and, on occasion, sang some
of his own risqué songs, accompanying himself on piano. Ban-

dini and Pee Wee collaborated at least once more, writing "You're There in a Dream," published by Standard Music Publishers, Ltd.

"Competition is keen between the groups," Mary wrote to the Simpsons. "It's a game but they ain't kiddin'. You know, who *is* the most popular and stuff . . . Muggsy was all over the joint last night selling albums. I'm afraid Russell was never cut out to be a super-salesman. But as for that Muggs, well, he does fine." Mary was correct. The competition among the men on the bandstand was nothing to kid about. They remained respectful of each other's talents; nevertheless there was a strong competitive spirit among them. "That whole gang of guys was very jealous of each other," clarinetist Kenny Davern said. "There was an animosity that was not far beneath the surface." And Pee Wee was usually the butt of the tension-relieving quips.

The gang at Nick's were between a rock and a hard place. The pressure wasn't coming just from Leonard Feather and the boppers: it was also coming from the purists, who rallied around the faltering sounds emanating from the newly purchased trumpet and teeth of Bunk Johnson. Johnson was one of the earliest figures in jazz who wrote his own version of jazz history by associating himself with the Buddy Bolden band and declaring himself to have been Louis Armstrong's teacher. With the boppers on the left and the purists on the right, the musicians of Pee Wee's generation were in the middle—not the real McCoy according to the Johnson camp, and too square for the followers of Charlie Parker. Although not evident at the time, those forces would squeeze many outstanding musicians—black and white— into critical limbo, Pee Wee among them. It would be nearly a decade before they were rediscovered. But at the beginning of 1945, Pee Wee and the other Nicksielanders were still optimistic. "We're working hard," he wrote to an English fan, Alistair Cameron. "Jazz is coming into its own here in the States. Being really lah-de-dah with concerts and the pink tea crowd discover-

ing an 'authentic American culture.' It is smart to like jazz.
Well, what the hell, we can eat now. Even the movies seem to be
on the verge of taking us up."

Pee Wee appeared on all four of the March broadcasts from
the Ritz Theatre. There was one more broadcast in the series,
on April 7, but Joe Dixon was the clarinetist. Forty-six half-
hour programs had been broadcast and preserved on 16-inch
radio transcriptions. This, along with the numerous recording
sessions over the previous sixteen months, accounts for more
than a third of all that has been preserved of Pee Wee's music
throughout his career. In contrast, there would be only three
formal recording sessions during the next eight years, and relatively few preserved broadcasts or concerts.

Wanda and Jerry Simpson sent Pee Wee a bottle to celebrate
his thirty-ninth birthday on March 27, 1945. Mary wrote:

> Your birthday drink came on his day off and I honorably (I open
> all the mail) gave it to him. We got good and drunk and I behaved
> badly. I decided to hear Dorsey and went uptown with a girl I
> picked up in a store and fortunately it was his night off too. Went
> back downtown, tore up Nick's, lost my wallet and Pee Wee tells
> me that the maid of a friend of mine brought me home. I don't
> know how *that* happened. I haven't called my friend and Pee Wee
> and I didn't talk to each other for a week.

The Russells' marriage continued to have its ups and downs. In
May, Mary wrote to the Simpsons,

> We're being angels. Saving our money and making marvelous
> plans about buying a station wagon and spending the rest of our
> life (or lives) in Mexico. Wanna come along for the ride? We'll lie
> in the sun and drink tequila and play records all day long forever.
> Or for a year or two at any rate. And we really mean it. We plan
> to do that in two or three years. Doesn't it sound wonderful?

During April, Ernest Anderson booked Monday night concerts at Philadelphia's Academy of Music and in Hartford, Connecticut, and Boston. Pee Wee joined Art Hodes for a college

concert in Connecticut during May. A Saturday afternoon concert from Nick's was broadcast nationally in June, but no recording is known to exist. Mary wrote of the broadcast to the Simpsons that "Pee Wee was fine. Best he ever played."

Mary and Pee Wee's life-style was beginning to erode their marriage. When one of them was on the wagon, the other was drunk. When they were both drinking, the fights intensified. Mary wrote to the Simpsons in July:

> I've got the blues tonight. Pee Wee and I have been indulging ourselves in a series of quarrels and now we've reached the silent stage, if you know what I mean. It hasn't been silent for the last couple of breaks. He's come up between sets to make up but I'm not having any. And he is drunker every time. I don't know what happens tomorrow. I may move except that it's almost impossible to get a place to live. I can't stay with my family because they are so fond of Pee Wee that I don't want them to know anything is going on. I have a number of friends I can stay with but I don't want them to know, either. So it all depends on an apartment. Or anyplace at all. I'll see what I can do tomorrow.

But those who knew the Russells knew that tomorrow would only bring more of the same—the volatile couple defiantly fending off the world.

On August 13, Mary wrote to the Simpsons again:

> Pee Wee quit his job and right now he is in bed trying to get in shape so we can go to the country. He's been sleeping, drinking egg nogs and shaking himself to death for the last 36 hours. He's damn near collapse and Nick has been pretty nasty to him. Always certain that Pee Wee will stick around. Anyway, I decided that I'm goddam sick and tired of his being constantly victimized. It didn't take much enouragement to make him walk out. Nick probably has a dirty trick or two left up his sleeve but Pee Wee covered himself pretty thoroughly.
>
> He quit in disgust with himself and Nick and being alive. He'd been blind drunk for two months before that. Reached the stage where he couldn't blow a note. Sick, drunk, afraid to face

reality and Nick taking advantage of him by bawling him out in
front of customers, and, incidentally, using me verbally as an
excuse for the way he was. There was one point where Pee Wee
went for his gun. Stuff like that.

So I made Pee Wee quit. He sent in two weeks pay for Boojie
Centobie (who is a swell guy) and his notice at the same time.

We're leaving for Westport, Connecticut, in a couple of days.
We'll stay a couple of weeks and I'm going to work on getting Pee
Wee back in shape. I don't know what Pee Wee is going to do
later and I care less. I want him to get well and not to worry
about work. We'll be okay.

When they came back to New York around the end of September, Mary again reported to the Simpsons:

That was fine stuff for Pee Wee. He ate three times a day,
stopped drinking and got stuff like sunshine and air. So he gained
fifteen pounds, got brown, lost the bags under his eyes. And I
became bitter and resentful. We were staying with a friend and
for three weeks I cleaned, cooked, shopped, washed dishes,
washed clothes, picked up after guys . . . and acted like a cop
to Pee Wee to keep his drinking at a minimum.

By the time we got back to New York and I made my mind up
to do that real quick, a matter of minutes, I was quivering like a
race horse at the post and all my good resolutions for the last six
months went down the drain. I got as drunk as a bastard and
stayed that way. Especially when everything went wrong. And
when I get drunk Wee gets twice as drunk and he can't pull
himself together. All that lovely health from the country is gone.
I've got to stay real sober most of the time to keep him close to
that. But when I'm drunk everything is bad. And at one point we
even clipped each other. Everything stinks.

Bob Mantler, who puts out one of those little jazz magazines,
is cleaning the house, trying to make us stop drinking and is
practically using forced feeding on us.

Fun things have happened too. Let me tell you about it. One
night, visiting a friend on Charles Street and a block party going
on under our noses and Pee Wee says wistfully, "That's Casey
playing on the street.' (Incidentally, Casey the bass quit Nicks

too.) So I said: "Aw, you're nuts, that's an electric guitar." So the Russ said: "I don't give a s--- if it's an ocarina, it's still Casey. I'd recognize his playing if he played a goddamed french horn."

So Russ went out on the street, found the three piece band, colored kids, playing for the block party. And Casey, who lives on Charles Street and can't ever resist music playing, you guessed it, borrowed an electric guitar. So Pee Wee went home, got his clarinet and they played their butts off. Someone called up someone and newspapermen came along and took pictures. A saxophone player came out of one of the houses. All the neighborhood kids stopped dancing and crowded around to listen and I never had more fun in my life. The boys finished a set in Nick's band and in the distance heard Wee and Casey, and Muggsy came hurrying along with his stuff. It was wonderful. Except that after that all the Welcome Home block parties tried to date Pee Wee and Casey.

Casey recalled the street party in a letter to Chicago record collector Jim Gordon. "The street dance emptied every joint in Greenwich Village," Casey wrote. "It sure was a good bash."

The gang at Nick's was finally breaking up. After many years of talk and numerous false starts, Condon decided to quit his meal ticket to go into the bar business himself. He credited Burgess Meredith with putting the idea into his head, but it was a natural for a promoter like Condon. Pete Pesci, the manager of Julius', the musicians' hangout around the corner from Nick's, provided the financial backing and assumed half-ownership. Actually Pesci ran the club and paid Condon a salary. They had hoped to open the club at the beginning of October, but ran into delays obtaining a liquor license. The lineup for the club's opening at first was to have been Max Kaminsky, Brad Gowans, Pee Wee, Bud Freeman who had just been mustered out of the armed forces, Joe Sullivan and Bob Casey.

Condon's club, at 47 West Third Street, was nearly hidden between two New York University buildings, in a space once known as the Greenwich Village Inn. The building had been constructed in the early part of the nineteenth century and has been described as "a building that could not have been built on purpose; it was impossible to identify its particular school of architecture but, if anything, it was a cross between a Greek temple and a Venetian *palazzo.*" Inside, tables were jammed into the area in front of the bandstand, which was draped in back and on the sides with bright yellow and reddish-orange striped fabric. A second-level balcony, running along the sides of the room, was also crowded with tiny tables. On December 20, 1945, Condon hung out the sign of the "pork chop," the club's logo, a cubist abstract design of a guitar by Stuart Davis, and watched the place fill up.

While awaiting Condon's opening, Pee Wee appeared in Boston on October 7, 1945, at an afternoon concert sponsored by the local jazz society. Billed as "The Thin Man of the Clarinet," Pee Wee was paired with "the trumpet sensation of the year," teenaged Johnny Windhurst, and accompanied by Boston's legendary traditional band, Charlie Vinal's Rhythm Kings. Then Max Kaminsky landed a job at the Copley Terrace in Boston and asked Pee Wee and Brad Gowans to join him. His band opened there on October 23. "We were on the air every night," Kaminsky recalled,

> so that people knew we were there and could hear what we could do and business began to pick up. Soon the place was doing so well that when Eddie Condon called up from New York to ask Pee Wee and Brad and me to come into his night club . . . I told him I couldn't come. Brad decided to accept Eddie's offer, but Pee Wee stayed on with me in Boston.

Kaminsky had been saving his paychecks at the Copley Terrace and had put together $3,000. He followed Condon's example with his own club in Boston. "Maxie's," in the basement at 220 Huntington Avenue, opened February 3, 1946. Kaminsky wrote:

> . . . When I was all ready to open, I discovered I had somehow spent all my money without being able to get a license to sell even Coca-Cola or a hot dog, let alone liquor. All I had was a cabaret license, and all I could provide was music, chairs and ashtrays. I opened anyway. Customers had to sit on wooden folding chairs and couldn't even order a glass of water.

Opening night the guest star was Joe Bushkin. For two weeks the band played for the few thirsty souls, including Red McKenzie, who stopped by. "Benny Goodman came in to see Pee Wee and offered him a job playing second tenor," Mary, who served as "assistant to all" at the club, wrote to the Simpsons. "He said when Pee Wee is good he's the best but when he is bad there is no one in the world as awful."

Kaminsky dissolved the enterprise and went back to New York to work at Condon's:

> This is one time Pee Wee and I have our train fare back to New York. I've shot my savings, and it was worth it for the music the boys played. It was worth it to discover how great all kinds of people can be when some good music hits them . . . I figured from the start that this band was my hobby. It was a hell of a hobby while it lasted.

Pee Wee rejoined the band at Nick's with Muggsy and Miff and played there through November. Not much had changed, except that the crowds had thinned out, thanks in part to the success of Condon's rival club.

The great stride pianist Dick Wellstood remembered a typical night at Nick's around this time:

> One of the most robust front lines ever—Muggsy Spanier, Pee Wee, Miff. The band was in a good mood that night. Muggsy, who had been a pitcher and who had a move like Whitey Ford's, had just thrown a whiskey-soaked napkin across the room and hit the bartender neatly with it just below the ear. Miff, being his usual glowering self, watched the fun behind steel-rimmed glasses, his tiny prison. Pee Wee leaned back, his drink under

his chair, rocking gently, and expelled such a cloud of spiky chains from his clarinet, so many blue wedges, sharp stars of sound, sharp-edged daisies of sound, that Muggsy turned and yelled, "Blow, Pee Wee, blow!" And Pee Wee closed his eyes and blew. And I was there.

One of Pee Wee's strongest supporters among the critics, Charles Edward Smith, was hired by Moe Asch to produce a series of albums for Asch's Disc Records. The band was basically the one at Nick's, with Vic Dickenson in the trombone chair and Nick's perennial intermission pianist, Cliff Jackson, at the keyboard. They were the first records to be issued under Pee Wee's name since Nick's Manhattans, more than a year earlier. The session produced an original blues by Pee Wee named after his boyhood home, Muskogee. He recorded another burning version of "Take Me to the Land of Jazz" and was featured for the second time as a vocalist, his hoarse voice croaking out the words in an amiable garble. In a *Down Beat* review of the album, the critic pointed out that "Charlie Smith remarks that Pee Wee sings just like he plays. If I were Pee Wee, I would smite him dead for that crack." Two tunes Pee Wee associated with Bix, "Since My Best Gal Turned Me Down" and "I'd Climb the Highest Mountain," were also included in the album. While Bix never recorded the latter, Pee Wee remembered his playing the tune over and over on the piano, exploring its peaks and valleys for hidden harmonies.

Nick Rongetti, Pee Wee's principal employer for more than a decade, died suddenly on July 25, 1946. Despite Nick's stormy relationships with his musicians, he was fondly remembered as a man who loved the music and supported it when it wasn't popular by hiring many who otherwise would have had to quit the music business. Art Hodes said that "the big thing he did was that he founded a home for jazz music during the dark ages, the lean years." Pee Wee said he was a "soul man, who ran the most friendly, cleanest little jazz room around." Mrs. Rongetti decided to keep the club going and managed to do so for several more years.

There was a brief flurry of excitement when, in August, it appeared that Mary was pregnant. Pee Wee "had the cigars all ready to pass out," he said. The couple had been trying to have a baby for over a year. Bassist Bill Crow reported that a sober Pee Wee arrived at Nick's

> in good focus, didn't drink all night, and actually held conversations with friends that he recognized. A couple of days later, when Mary found out her pregnancy was a false alarm, Pee Wee returned to his routine, arrived at work in an alcoholic fog, speaking to no one, alternately playing and drinking all night long.

In September, Charles Edward Smith produced another four-record album for Disc Records, this time under Spanier's name. Pee Wee was featured on another blues vocal similar to the one they had cut for V-Disc more than a year before, "Pee Wee Speaks." This time, it was called "Pee Wee Squawks." It was Pee Wee's third vocal in a year—and, thankfully, his last.

A few weeks after the session, Pee Wee played a Town Hall concert, sponsored by Bob Maltz' New York Jazz Society. The musicians included Spanier, Mole, Art Hodes, Pops Foster, George Wettling, Wild Bill Davison, Johnny Windhurst, Vernon Brown, Mezz Mezzrow, James P. Johnson, Baby Dodds, Sidney Bechet and many others. But by late 1946, such events were beginning to be passé.

A review of a night at Nick's by critic Bob Aurthur appeared in Art Hodes' *Jazz Record* in November:

> No jazz band should stay in one spot for more than a certain period of time . . . shall we say, arbitrarily, eight weeks? At Nick's, we find a band that, on paper, shapes up as one of the best in the land but actually isn't doing too much. Pee Wee

Russell, the nonpareil, plays clarinet at Nick's, but he is certainly not the Pee Wee of old who used to play such tremendous clarinet that you would want to cry out of sheer joy. No, Pee Wee is now a victim of too much publicity, too many jokes at his expense, and maybe, as I suggested before, too much clarinet playing in one place.

Now Pee Wee honks, wheezes and squeaks, and even uses a mute to achieve an effect . . . plaudits from the crowd . . . that he used to get by playing good clarinet. But both Pee Wee and the crowd have changed, a change for the worse, of course. Anyway, the people who come to Nick's talk so loudly during the set that you can't hear Pee Wee, so it doesn't make much difference.

It was time to move on, and Pee Wee knew it. The final straw was a conflict over the Manhattan records. As *Down Beat* reported:

In a hassle between Pee Wee Russell and Nick's, the famed clarinetist decided to remove himself from the fixtures at the jazz spot where he has been on display more or less regularly for the past ten years.

Crux of the differences with Mrs. Nick Rongetti, widow and successor of the club's founder, was a recording contract held by the Manhattan Music Corporation, a Rongetti enterprise. Pee Wee claims that Muggsy Spanier, Miff Mole and Pee Wee each had a year's contract with Manhattan to make records that were, until recently, to be sold only at Nick's. Terms prevented the artists from recording elsewhere but, says Pee Wee, guaranteed $2,000 to each of the three, less $100 in various charges.

"In the past sixteen months—four months beyond the contract date—I've been given only $108," the article quoted Pee Wee. But Jack Russell, the manager at Nick's branded the whole story a lie and said that the money Pee Wee was to get from records, "something closer to $1,200," was getting to him in proper fashion. The article concluded by noting that Pee Wee had joined Eddie Condon's, where "he's long had a standing offer." Pee

Wee and Condon, according to *Down Beat,* were "known to be old-line feuders. But Russell expects no undue stress. 'After all, I can always talk to [drummer] Dave Tough,' he said."

Pee Wee's conversations with Tough didn't last very long, however. Writer Bill Gottlieb interviewed Tough in the September 23, 1946, *Down Beat,* and the article ("Dixieland Nowhere Says Dave Tough") set off an uproar. "Things were going too well at the sign of the distorted pork chop," wrote Bob Aurthur in *Jazz Record.* "Davy Tough ('the Dizzy Gillespie of the drums'—Brunis) had written an article in *Down Beat* claiming that Dixieland was dead and rebop had definitely taken its place. Condon's, said Davy, was leading the way.

> A step in the right direction was taken when Pee Wee Russell changed sides and came from Nick's into Condon's. Now Pee Wee isn't playing half what he used to, but the half he is playing is still plenty good enough . . . Tuesday night is about the best time to hit Eddie's for it's on that night that you can catch the weekly "Ham Session." Brunis is usually there. So is Wettling, Hackett and, best of all, Sidney Bechet, and anyone else who happens to be in town.

The band was split by the bitter dispute between Tough and Brunis, who had replaced Ohms. "Citing Georg Brunis' propensity for playing trombone with his foot and adding that when he plays legitimately it still sounds the same, drummer Dave Tough left Eddie Condon's shortly after Brunis entered on a full-time basis," *Down Beat* reported:

> Midst bad blood that could be smelled all the way over to Nick's, there was a changing of the guard at the Greenwich Village club on January 30, with Brunis, Wild Bill Davison, George Wettling, Morey Feld, Sid Weiss and Al Hall replacing Tough, Lesberg, Max Kaminsky and Freddie Ohms.
> Bud Freeman, who was also recently added to the Condon crew and who is shortly scheduled to leave for Brazil, isn't involved in the hassle, nor is the ever-present Pee Wee Russell.
> According to Tough, business at Condon's as elsewhere

wasn't at old levels. Condon's silent partners "put Eddie over the barrel and made him make changes."

Eddie brought in Brunis and Davison, easing them in via a Tuesday night jam session turn, then adding them full time when the sessions pulled full houses. This soon put Maxie and Freddie on the outside. Dave and Jack Lesberg joined the pair, refusing to work with Brunis or Davison, who Dave considers, respectively, "a clown and a musical gauleiter."

The departing barefoot boys were involved in a *cause célèbre* on *Down Beat*'s pages some months ago when they struck out against the "antiquated dixieland" of their boss, Eddie Condon.

"The Beat articles," [said Tough] "didn't create any new difficulties. It simply brought them to a head, gave the issues a 'name,' made the arguments black and white. As a result, we had to give up most of the gains we made towards instilling modern music into Condon's routine. But Eddie is a good guy to work for; and things were satisfactory until these new dead-jazz characters were added. To keep up with them, I'd have to learn to drum while standing on my head."

Although his sympathies were with Tough, Pee Wee stayed aloof from the fray. Several of the Condon band were recruited by producer Al Rose around this time to play at Blair House for President Harry Truman. The band flew to Washington, where they played for an enthusiastic Truman. The President even replaced pianist James P. Johnson on several tunes, including "Kansas City Kitty," much to the delight of himself and most of those present. The drummer, Baby Dodds, who had no idea who Truman was, remarked on the way back to New York, "that little fellow with the glasses couldn't play much piano."

Ernest Anderson had departed for England, where he spent the next couple of decades, leaving Condon to fend for himself. Always an able promoter, the new club owner swung into action. His annual winter concert series at Town Hall had already started, and continued through April on the first Saturday of each month. Pee Wee and other Condonites appeared at a jazz festival sponsored and broadcast by WNYC in February.

Although Commodore Records was temporarily inactive, Milt Gabler was producing records for Decca. One of the first sessions he did for that company was an all-star Condon date on August 6, 1947, with a front line of Jack Teagarden, Wild Bill Davison and Pee Wee.

Condon was also renewing his television career—on hold since the experimental programs in 1942—with his "Floor Show" programs on WPIX-TV. Now, five years later, television was still a novelty, but more and more sets were being sold, at least in major metropolitan centers where commercial stations were broadcasting for several hours every day. Condon's show was a mixture of his brand of jazz, popular singers (Rosemary Clooney was a regular) and—to take advantage of the visual aspects of the medium—tap dancers. Witty repartee was supplied by Condon and the show's announcer, Carl Reiner. Pee Wee was not featured on most of the programs, the clarinet chores usually being handled by Peanuts Hucko.

Pee Wee starred in another of Bob Maltz' jazz concerts on November 29, 1947. Early in 1948, Pee Wee returned to Nick's for the last time, as a member of Billy Butterfield's band. Fred Ohms was on trombone, Jess Stacy on piano, Bob Casey on bass and Joe Grauso on drums. Stacy remembers:

> I played with Pee Wee and Billy at Nick's. Pee Wee was such a nice, lovable guy he'd go and sit at the tables with fans. When they'd ask him what he wanted to drink, he'd just order what the host was having. So he'd mix his drinks up all night. He'd drink anything. He held it well, though. Never got nasty or anything. He was a good drunk.

Despite the gig at Nick's, Condon's was now Pee Wee's home base, though he took frequent leaves of absence to appear with other bands around the country. In June, 1948, he appeared at the Ken Club in Boston with Max Kaminsky's band, which included Miff Mote, and George Wein, piano. It was Pee Wee's first gig with Wein, who would later have a major impact

on his career. "It was my first job with name musicians," Wein
said. "I had played around Boston with Charlie Vinal's Rhythm
Kings and I learned all the dixieland tunes they were playing.
Miff Mole was absolutely wonderful, and that's when I found
out what Pee Wee meant. I really got to love Pee Wee at that
time."

In October, Pee Wee opened at the Blue Note in Chicago
with Muggsy Spanier and Miff Mole. Art Tatum was the intermission pianist. On October 16, Pee Wee appeared at a Civic
Opera House concert, m.c.'d by Dave Garroway and including
the Spanier band, the Art Van Damme combo, Art Tatum and
Herb Jeffries. "Russell's not-quite-there clarinet" and the Spanier band, reported *Down Beat,* "got the biggest hand."

On October 30, 1948, he appeared at a concert billed as
"Journey into Jazz," promoted by Al Rose and staged at the
Academy of Music in Philadelphia. The bill featured "30 star
jazzmen," including Wild Bill Davison, Sidney Bechet, Joe Sul-
livan, Muggsy Spanier, Bertha "Chippie" Hill, Pops Foster and
Baby Dodds.

Condon used Pee Wee on one of the Floor Show telecasts on
December 16 and January 15, 1949. Gowans lined up a series of
Sunday afternoon dates for the Nassau Jazz Society at Jack
Fowler's, near Trenton, New Jersey, for the winter season.
Gowans' band included Bobby Hackett and Pee Wee.

By February, 1949, Pee Wee was playing in a trio with Art
Hodes and Herb Ward, bass, in the backroom at a new club, the
Riviera, located on Seventh Avenue near Nick's. "Our back-
room thrived and jumped," Hodes remembered. Jack Teagarden
sat in for a week, and other jammers included clarinetist Tony
Parenti, George Wettling, Hot Lips Page, one of the original
Mound City Blue Blowers, "Josh" Billings, on suitcase, and
blues belter Chippie Hill. The group lasted until May, when
Hodes returned to Chicago. Pee Wee went to Chicago too, and
was featured with Hodes' "Back Room Boys"—on May 16 at the
Silhouette Club. Brownie McGhee and Mama and Jimmy

Yancey rounded out the bill. After the concert, Pee Wee left Chicago as quickly as possible and returned to the "Backroom" in New York. Hodes briefly returned to New York to play solo at the Village Grove. He was replaced by Willie "the Lion" Smith in Pee Wee's trio.

In July, Pee Wee and a large group of New York musicians descended on the rural community of Stockertown, Pennsylvania, to play at the wedding of a close childhood friend, Helen "Daisy" O'Brien, and Duane Decker. The musicians included Bobby Hackett, George Wettling, Ralph Sutton, Jonah Jones, Willie "the Lion" Smith and several others. Daisy, as she was known to her many friends, was the photo editor for the *Ladies Home Journal*. She arranged to have Charles Peterson photograph the wedding. In fact, the event took place on Peterson's 16-acre farm. The groom was a sports novelist and also a jazz fan. "The large congregation of famous musicians fresh from Nick's, Condon's, Ryan's, Savoy, Onyx, and so on, played not only the Wedding March, but also predictably 'A Good Man Is Hard To Find' and many other jazz standards," recalled Charles Peterson's son Don. "The music gang looked like a bunch of Al Capone characters arriving in their dark suits from New York in this little rural community. At one point, Helen accompanied Pee Wee on suitcase drums."

In August, 1949, Pee Wee appeared again on Condon's pioneering television program, Floor Show, which was celebrating its first anniversary on NBC (it had originated on the Dumont network). The guest of honor was Louis Armstrong, and the resulting broadcast included a rousing performance of "We Called It Music," a plug for Condon's book of the same title, on which the raconteur strummed a rare guitar solo, and "Chinatown," on which Pee Wee was barely able to rise to the occasion. They are the only performances known to have been preserved of Pee Wee playing with Armstrong.

During October, Pee Wee appeared at several concerts promoted by Bob Maltz at Central Plaza. On October 7, the bill included Bobby Hackett, Billy Butterfield, Sandy Williams, Joe Sullivan, Art Trappier, Art Hodes, Chippie Hill, James P. Johnson, Max Kaminsky and Pee Wee. On October 21, 1949, he appeared at another Maltz concert with Yank Lawson, Hackett, Frank Orchard, Johnson, Sullivan, Baby Dodds and George Wettling. Woody Guthrie was the special guest star.

When Hodes was hired to play at the Blue Note in Chicago for eleven weeks that winter, Pee Wee went with him, but without Mary. His drinking had been as bad as ever and his behavior even more erratic. The split came as no surprise to friends and acquaintances, since they seemed always to be carrying on a boisterous public feud. Even though Mary drank to what some would consider excess, she nevertheless tried to help Pee Wee with his more serious, even life-threatening, problem. "When Jimmy [McPartland] was drinking," said Marian McPartland, who was married to the trumpeter, "I always hated anyone who gave him a drink. And I know Mary felt the same way concerning Pee Wee. Buddies would say, 'Have one, you're not an alcoholic.' The pressure to drink was great." Years later, Mary told Whitney Balliett:

> Once when Pee Wee had left me and was in Chicago, he came back to New York for a couple of days. He denied it. He doesn't remember it. He went to the night club where I was working as a hat-check girl and asked to see me. I said no. The boss's wife went out and took one look at him and came back and said, "At least go out and talk to him. He's pathetic. Even his feet look sad."

The break appeared to be permanent.

Without Mary around to lean on, Pee Wee's health began to decline precipitously. Hodes and the other musicians around Pee Wee became concerned. The frail six-footer, who had weighed as much as 144 pounds a few years earlier, had stopped eating and lost at least twenty pounds. "How vividly I recall McPartland visiting us and taking in the 'state' of Pee Wee's health," Hodes said, "insisting that Frank Holzfiend, the manager, let us put him in a hospital for some treatment and rest. Frank was with us, but thank God for Pee Wee's strength. The 'Indestructible One' came back, as he continued to do down to his final appearance."

In February, 1950, Hodes' band, with Pee Wee and Lee Collins, trumpet, was renewed for another fifteen weeks. Singers Chippie Hill and Sarah Vaughan were also on the bill. In mid-February, Vaughan was replaced by pianist Erroll Garner. In April, Pee Wee moved over to the Brass Rail with Georg Brunis, who had put together a band of local Chicago musicians. Pee Wee was described by those who saw him during this time as even more nervous than usual, drinking very heavily, emaciated and frequently ill. After all the years at the top of the jazz polls and having the pick of the top jazz jobs in New York, Pee Wee found himself and his career on a downhill slide, ignored by the jazz press and fans alike. He was a has-been, but that fate was shared by many once-famous jazz musicians during the late 1940s. All of the various manifestations of jazz were having a rough time finding venues. Bop had failed to attract a mass audience the way the swing bands had, and dixieland, Nicksieland or Condon-style had been branded as corny, hopelessly out of step with the times. Furthermore, the night-club business was in a slump as more and more people stayed home to watch television.

Pee Wee, in his feeble mental and physical condition, took the rejection personally. When Brad Gowans stopped by Chicago on his way to California and suggested that Pee Wee go with him, it seemed to be just what the doctor ordered. The change of scenery and mild climate might help him regain his equilibrium. And Gowans had expectations of a job there when they arrived. In the spring of 1950, they set off together in a 1920 Rolls-Royce that Gowans, an automobile buff, had managed to obtain and rebuild. The trip to the coast was precarious, with much happy tinkering by Gowans along the way. They finally reached San Francisco in July. There they found Ralph

Sutton, a young pianist from St. Louis who had been featured on several of Rudi Blesh's This Is Jazz programs in 1947, and had made his first solo records a year earlier. Pee Wee opened at the Say When in San Francisco with Sutton's band for three weeks. "He was pretty weak then," Sutton recalled. "We had a one night gig in Sacramento, and Pee Wee was so sick he couldn't make it, so Ed Hall filled in."

A few weeks later, Pee Wee was able to play a gig in Sacramento. Trumpeter Marty Marsala, who had settled in California, added Pee Wee to his band for a two-week engagement at the Clayton Club in the state capital. After that, Marsala and Pee Wee gigged around central and northern California for a few months, and in the fall, returned to San Francisco and worked with a local group which included Slim Evans, tenor; Johnny Wittwer, piano; Pat Patton, bass; and Chris Krider, drums, at a run-down 1920s speakeasy, Coffee Dan's. While in San Francisco, Pee Wee stayed with Patsy and Pat Patton, who looked after him. The gaunt clarinetist's body was frequently wracked with pain. The Pattons did their best to care for him, but he was unable to eat even his usual dinner of canned tomatoes and keep them down. Most of his nourishment came from the milk he consumed with gin. As sick as he was, he could not stop playing. He resisted attempts to get him to check into a hospital, in part because he feared people would think he could not play anymore. He was falling into a deep well while maintaining all the way down that everything was fine. Nevertheless, more and more frequently he had to sit out a set or two. Some nights he was unable to make it to the club at all. The 44-year-old man was falling apart physically and mentally. On New Year's Eve 1950, he collapsed on the bandstand and was rushed to San Francisco County Hospital in critical condition. Because he had no insurance and no money, he was admitted to the charity ward. Although seemingly forgotten by the jazz world, Pee Wee was still enough of a name to warrant a United Press wire story on January 8, 1951, which reported he had been stricken in San Francisco with a liver ailment.

Someone brought the wire story to the attention of Helen "Daisy" Decker, the childhood friend whose wedding Pee Wee had played less than eighteen months before. She rushed into action in New York, where she was working as the photo editor at *Cosmopolitan*. With her contacts in the media, she was able to arouse interest in helping Pee Wee. A *Time* office memorandum revealed that Pee Wee was "broke and dying—down to 73 pounds," and reported to its readers that "his liver [was] almost gone from lost weekends." Daisy also made the rounds of jazz fans and friends in New York, raising money so that Pee Wee could be removed from the charity ward and die with some dignity. The party that had started thirty years before in the bands that played the riverboats on the Arkansas River was over.

10 • • • •

Rebirth

The telephone rang in Mary Russell's small apartment. She heard the sob-choked voice of Eddie Condon. "Our boy is going tonight," he said. Such was the prognosis. The doctors held out little hope for Pee Wee's survival. The diagnosis was acute pancreatitis, an often fatal illness associated with chronic alcoholism. On January 8, 1951, the Associated Press carried a report that he was in need of blood donations. "Russell has already been given one pint and may need more transfusions in order to survive. A physician says there is no more blood available for Russell at the hospital." His face was sunken; huge boils had erupted on his neck and his weight had dropped to 73 pounds. The long, thin fingers that had wrung such passion from the clarinet for more than thirty years were little more than skin-covered bones. Nevertheless, he managed to hang on while doctors attempted to stabilize his condition enough to perform an operation. For several weeks, Pee Wee barely seemed to be holding his own.

At first, word of his hospitalization was slow getting around despite the best efforts of Daisy Decker. But once the jazz world learned of his situation, an amazing thing happened, amazing at least to Pee Wee: musicians and fans who he was sure had forgotten him—many he never knew—came to his aid. Jack Teagarden and Louis Armstrong were among the first. The two were in San Francisco, performing to turn-away crowds at Club 150. The great trumpeter told the press he wanted to "play one for Pee Wee," and on January 22, his All Stars headlined a benefit at Doc Dougherty's Hangover Club. Marty Marsala's band, Brad Gowans, Meade Lux Lewis, Mary Ann McCall, the Walter Mitchell Trio, Tut Soper, Smokey Stover, Pat Patton, Dorothy Bennett, Albert Nicholas and Julian Lane were among the musicians who performed. Ralph J. Gleason reported in *Down Beat* that the musicians "played to a sardine-can-packed house till the wee hours. By mid-evening the Standing Room Only sign was out, and Doc had to keep people on the sidewalk until someone left. And few left." It was "one of the greatest tributes to a jazzman this country has ever seen," Gleason concluded. A half-hour of the benefit was broadcast on KNBC with Jimmy Lyons announcing.

The benefit raised more than $1,500, which enabled Pee Wee to be moved from the charity ward of San Francisco Hospital to a private room at Franklin Hospital. As important as the more comfortable quarters were, the support of the musicians and fans meant even more. It lifted his spirits from despair to hope. He felt, perhaps for the first time in his life, that his peers genuinely liked him and respected his talents. The realization came as quite a surprise to him. On January 29, 1951, he underwent major surgery. "They had him open like a canoe!" Eddie Condon reported to his friends. During the course of the operation, more than twenty cysts were removed from his liver and the doctors also discovered the reason why Pee Wee could not consume solid foods: a birth defect that impeded the passage of solid food.

> All during the forties, I'd be hungry and take a couple of bites of delicious steak [Pee Wee told Whitney Balliett] and have to put the fork down—finished. I lived on brandy milkshakes and scrambled-egg sandwiches. And on whiskey. The doctors

couldn't find a thing. No tumors, no ulcers. I got as thin as a lamppost and so weak I had to drink half a pint of whiskey in the morning before I could get out of bed. It began to affect my mind, and sometime in 1948 I left Mary and went to Chicago.

Everything there is a blank, except what people have told me since. They say I did things that were unheard of, they were so wild. Early in 1950, I went to San Francisco. By this time my stomach was bloated and I was so feeble I remember someone pushing me up Bush Street and me stopping to put my arms around each telegraph pole to rest. I guess I was dying.

Today alcoholism is known to be a deadly disease; a predisposition to it may even be linked to gene structure. But in 1951, it was generally regarded as a character flaw, a lack of will power. Pee Wee, like Bix, was physically and psychologically addicted to alcohol from his early teenage years. It is possible that Pee Wee's nervous twitches and "shakes," already evident in his late teens, were caused by neurological damage, the result of imbibing bad home brew.

Early jazz historians portrayed the premature deaths of Bix, Bunny Berigan, Fats Waller and the others in very romantic terms, casting the doomed jazzmen as misunderstood poets whose self-destructiveness was a reaction to an uncaring world. Some writers, for example, maintained that Bix drank himself to death at age 28 because of the boredom of playing the turgid arrangements of the Paul Whiteman orchestra. But two-fisted drinking was the ethos of the times; Pee Wee's affliction was all too common. During the twenties and thirties, many jazzmen felt a special obligation to maintain their hard-drinking reputation. Most would deny they were alcoholics—that was a term reserved for skid row bums—and maintain they drank no more than others in their circle. They often were encouraged to drink on the job by employers and fans. Their places of employment invariably were speakeasies, and their job, as far as the club owners were concerned, was to keep the customers in a happy, drinking frame of mind. There was a feeling among some fans that musicians performed better when they were inebriated: the lessening of inhibitions would result in ever wilder improvisations. Fans plied the musicians with drinks and if a drink was refused, the musician was considered a snob or worse. Max Kaminsky recalled:

> Pee Wee liked to go to a bar, give the sign to the bartender, about three inches, put his foot on the bar and drink. He liked Rye and Vodka. He drank too much all his life. He was sick all the time. Every time he got really sick, he'd swear off and spend about a week or two on the wagon. I remember when we worked together on the stand he'd play beautifully, much better than some of the stuff he'd do under the influence. To me he sounded better. He had hallucinations when he'd drink.

The February 12, 1951, issue of *Life* carried a story on the clarinetist's plight. "No Sad Songs for Pee Wee" included photos showing his emaciated body with Louis Armstrong and Jack Teagarden hovering solicitously over him. "Through the small, ardent world of music lovers who like their jazz hot, original and unfettered by the 'paper' arrangements of the name bands went sad, long-expected news: cadaverous 'Pee Wee' Russell had caved in." The article continued in the past tense:

> Pee Wee was an unostentatious but emotional man, who kept his feelings bottled inside him until he put a reed to his lips. Then, in contrast to the suave glibness of a Benny Goodman, the emotion came eloquently out in rhythmic swirls of low register growls and mercurial high-note figures. Pee Wee never blew a note he did not feel, and he made others feel it too.

On February 21, a second benefit was held, this one in New York under the auspices of Eddie Condon at Town Hall, the scene of so many of Pee Wee's triumphs a decade before. Wild Bill Davison, Ed Hall, Cutty Cutshall, Joe Sullivan, Willie "the Lion" Smith, Gene Schroeder, Bob Casey, Buzzy Drootin, Ralph Sutton, Joe Bushkin, Jimmy Archey, Lou McGarity, Frank Orchard, Vernon Brown, Billy Butterfield, Henry "Red"

Allen, Bud Freeman, Ernie Caceres, George Wettling, Ray McKinley, Al Hall, bassist Irving Manning, Pee Wee Irwin's band, Lee Wiley and Peanuts Hucko were among the participants. Condon's many media contacts were pressed into service to ensure the success of the affair. "Pee Wee nearly died from too much living," Condon told the press. Newspapers contributed free advertising for the concert, disc jockeys plugged it and NBC offered free announcements. The benefit raised more than $3,000. A third benefit, smaller in scale but no less deeply felt, was held in Chicago on March 4 with Art Hodes, Georg Brunis and pianist George Zack among the musicians present. Pee Wee received more than $4,500 from the proceeds of the three benefits, a sizable total for the day. Reporting on the concerts, syndicated columnist Robert C. Ruark wrote:

> A great many musicians burn freely from both ends, like candles, and it is generally accepted that a great many wind up sick, sore, destitute or dead. Raw gin and irregular hours never stored up a vitamin surplus. . . . Pee Wee Russell, the professor emeritus of clarinet, seemed destined to die the other day, until the pleasure he had given the jazz world redounded to his benefit . . .

At first, it appeared that the operation was too much for Pee Wee's shattered constitution, but ever so slowly, there were signs of improvement. Finally able to eat solid food, he began to gain weight. The enforced abstinence from alcohol was clearing his mind. Despite the years of abuse and against all expectations, "The Indestructible," as Art Hodes had called him, began to recover. Still somewhat feeble and very underweight, Pee Wee was released from Franklin Hospital on February 27, less than a month after the operation. "I was still crazy," Pee Wee recalled. "I told them Mary was after me for money. Hell, she was back in New York, minding her own business." Upon his release, Pee Wee returned to New York, where Daisy and Duane Decker were prepared to care for him while he recovered. He was driven to the airport in an ambulance, accompanied by Fred Wyatt, a San Francisco newspaperman who had been among the first to tell the world about Pee Wee's illness.

Pee Wee arrived at LaGuardia Airport, bundled in blankets against the February cold, where he was met by Eddie Condon, Ralph Sutton (who had been hired as intermission pianist at Condon's club immediately after the engagement in San Francisco with Pee Wee), photographer Charles Peterson, Bob Casey and Buzzy Drootin. "A bunch of us stayed up," Sutton recalled. "We were all down at Eddie Condon's. We went out to Josh Billings' house, then we all went to LaGuardia. We took him down to Gramercy Park, where he stayed with Duane and Daisy Decker for a few days." After that, Pee Wee checked in at St. Clare's Hospital for tests. He registered under an assumed name, "so Mary couldn't find me," he later said. After his release from the hospital, he moved in with the Condons. According to Phyllis Condon:

> He recuperated here for a month. He was so ill his bed had to be changed all the time. There was some kind of draining from the operation. He had a hiatus hernia. When he was here he was the most quiet patient you could imagine. The room was always kept dark. He asked for nothing.

The doctors had warned Pee Wee that any consumption of alcohol would be fatal, and, for a while, Pee Wee was able to stay on the wagon. Soon, however, he began imbibing ale, and every now and then, a drink or two of liquor.

Mary sent him a dixieland fake-book for his 45th birthday on March 27, and Pee Wee called her at Tony Pastor's club in the Village, where she was working as a hat-check girl. "This is a new Pee Wee," Mary told jazz writer Barth Mackey in an extensive interview.

> I'm not sure I know him. For one thing, he talks a lot. He never used to. It's as if he were trying to catch up. And you know what? He told me not to swear! I got going about something and he

stopped me. "I don't want my wife to swear!" he said. Why, that isn't like the old Pee Wee at all!

What was the old Pee Wee like? the interviewer asked:

Well, nervous for one thing, and sentimental and kind of shy. He liked to be alone. You know that garbled way he talks to people so that no one can understand him? I think that's on purpose. At least he never did it with me. It's his way of avoiding people. Musicians get an awful lot of silly requests when they're playing—waltzes, tangos and stuff like that, and then there's always some collector who wants him to play a chorus from some tune he recorded twenty years ago that he doesn't even remember. Well, when he talked like that he gets rid of them. Pee Wee's a very intelligent guy. A nice guy too, I guess. That is, sometimes.

Mackey asked Mary what Pee Wee had been like at home.

He liked to listen to the ball games, and he never missed the Lone Ranger. He never stayed out late with the boys either. He was always home by 4 a.m. Then he'd go for a walk in the park real early in the morning with Nini. That's his dog. But he really loves the clarinet the best, and that's as it should be. After that he loves his dog. I guess I was third.

"There are lots of things that people don't know about Pee Wee," Mary continued. "One thing that they don't know is that he likes chamber music. He likes bop too, and he listens to Gillespie. He hasn't any of his own records, but he does have some Jelly Roll Morton's Red Hot Peppers. [Omer] Simeon is his favorite clarinetist. He asked everyone where Simeon was, and finally found him in a recording studio. By that time he was so excited that when they met he couldn't just shake hands with him, he had to throw his arms around Simeon and hug him.

But Pee Wee has changed. I'm not sure that I know him. He asked me if I were getting enough to eat. He never used to ask questions like that. And he doesn't want me to swear! I said: "Pee Wee, we're still separated you know," and he said: "Yeah, by about forty blocks." I really don't know what to do.

"Mary heard where I was," Pee Wee recalled simply. "She came over and we went out and sat in Washington Square Park. Then she took me home. After three years!"

His life was beginning to come back together. He resumed playing the clarinet, and on July 17, 1951, he felt strong enough to sit in with the Condon band on jam night. It was a different man on the bandstand than before. Looking back over his years at Nick's and Condon's, Pee Wee realized that he had

> a sorrow about that time. Those guys made a joke of me, a clown, and I let myself be treated that way because I was afraid. I didn't know where else to go, where to take refuge. I'm not sure how all of us feel about each other now, though we're "Hello, Pee Wee," "Hello, Eddie," and all that. Since my sickness, Mary has given me confidence . . .

For the first time in his life, Pee Wee was confident enough to organize his own band. He had fronted a band at the Little Club in 1938, but that band had been organized by Ernest Anderson. This band would be Pee Wee's band with the musicians he wanted, playing the music the way he wanted it played. He set about looking for young musicians—new faces, new ideas—for his band. The national publicity about his illness, the benefits and his miraculous recovery were still fresh enough to ensure some initial bookings. In the meantime, he was a regular at the weekend bashes in the Stuyvesant Casino, located at Second Avenue and Ninth Street. Plans for his own band, however, continued. Trumpeter John Dengler and trombonist Ephy Resnick had been playing together for a while in the Poconos and auditioned as a brass team. "The audition was held at Lou Terrasi's on 47th Street," Dengler remembered. "We tried out against trumpeter Bill Price and trombonist Dick Fails from Long Island. Bill Gammie, who was managing Pee Wee's fate at the time, said to me, 'You aren't a cornet player, but you are a musician,' and Ephy and I got the job." Teddy Roy, piano; Irv Manning, bass, and Eddie Phyfe, drums, completed the band.

Pee Wee called his band members, most half his age, his "kiddies."

Gammie acted more as a fan than a manager. Mary traveled with the band and handled the business. Their first gig was at the staid National Arts Club, 15 Gramercy Park South, on October 6, 1951. For the next couple of weeks the band rehearsed at Condon's, and then they drove to Aurora, Colorado, on the outskirts of Denver, where they opened for a three-week engagement at Bob Cummings' Zanzabar. Kenny John replaced Ed Phyfe on drums after Pee Wee turned to Phyfe during a set and muttered, "If anybody's going to change the tempo, it's me!"

There were other problems with the band. Dengler, who had never played at such a high altitude, was having trouble with his embouchure. "I blew out my chops on the first night," Dengler said. Attendance at first was very sparse. Zanzabar owner Cummings was not happy. "He arbitrarily cut Pee Wee's salary," Dengler recalled, "and Pee Wee did not have it in him to fight it. It was beneath him. So Bill Gammie decided not to take his cut—that doesn't sound much like a manager."

Irv Manning remembered the last Saturday night of the engagement. The club was virtually empty when "all of a sudden, in walked a whole crowd of people including Robert Mitchum."

> He took the whole crew of the movie he was making into the Zanzabar because he heard that his friend, Pee Wee, was playing there. We played an extra hour for him and his entourage. After the place closed we went back to Mitchum's hotel and he kept us up telling jokes until 3:30 in the morning. When we got back, Mary beat the hell out of Pee Wee because she thought he was running around. He didn't have the strength to cat around even if he wanted to. The next day, we all went down and played for Mitchum and the studio people at their location in Colorado Springs.

The band took the train to their next job, the Terrace in East
St. Louis, Illinois. Pee Wee made no attempt to contact any of
his relatives across the river, nor they him. Manning:

> We played behind the bar, and you had to go three or four steps
> above floor level for the bandstand. Pee Wee had only a narrow
> path to get around the drum set, so he would stick his finger on
> the cymbal to steady himself, which, of course, was of absolutely
> no use to him whatsoever. It was really very funny to see him
> walking his way past the drum set with one finger on the cymbal.

Even though the band was beginning to sound "real great,"
according to Manning, the engagement was a near-disaster. A
fire at the club, caused by defective wiring, closed them out a
few days early. Manning's bass, made in 1695, was singed. "It
killed the bass," Manning said. "The fire dried out all the sap in
the wood and it lost its vibrancy." The band limped on to the
next job at the Capitol Lounge in Chicago. Red Richards took
over as pianist for the engagement, and Ruby Braff, a 25-year-
old cornetist from Boston, replaced Dengler. "I knew Pee Wee
since I was 16 or 17," Braff said. "I used to go down to New York
for the weekend, and the first place I'd go was Nick's."

Pee Wee returned to New York for a week around the beginning of 1952, sitting in with George Wettling's Stompers at the
Stuyvesant Casino. Then he and his band were off to Boston for
a month's engagement. George Wein, the pianist who played in
Max Kaminsky's band with Pee Wee in June, 1948, had opened
his own club, Storyville, in the Copley Square Hotel. Pee Wee's
band was one of the first he hired to play there. The band's
regular broadcasts over WMEX attracted large and enthusiastic
crowds to the club. In addition to Pee Wee's improvisations,
Braff's cornet work was a special favorite with the crowds.
Wrote critic Nat Hentoff in *Down Beat:*

> No longer the mildly vague "character" of jazz lore, Charles
> Ellsworth Russell is fiercely proud of his new band and is deter-

mined to keep it together as long as he can. The combo, an unusual mixture of eclectic styles, is already a remarkably well-integrated unit, personally as well as musically, though it's been together only a few weeks. It demonstrated during its recent Storyville date here that Dixieland needn't be limited to static, repetitive figures over a prematurely senile beat.

When the engagement was over, however, jobs were not immediately available for the band. It had been the wrong time to start a band, publicity or no publicity. There were few clubs that employed musicians who played a non-categorical jazz. And Pee Wee found that dealing with the hassles of being a leader was not for him. The experience, however, did initiate Pee Wee's enduring friendship with Braff.

Pee Wee returned to New York in February, 1952, and resumed playing regular weekend gigs at the Stuyvesant Casino with George Wettling's band. In March, Pee Wee and Braff formed a new band. They opened at the Colonial Tavern in Toronto on March 31, 1952, and were held over through April 7. On April 12, Pee Wee was back in New York to play a concert at Town Hall with Bobby Hackett, Wild Bill Davison, Brunis, Jimmy Archey, Edmond Hall, Bud Freeman, George Wettling, Gene Sedric, Joe Bushkin, Lee Wiley and others. It was just like one of the giant "clambakes" of yesteryear.

Pee Wee's band opened at the Blue Note in Chicago on April 18 for a two-week engagement. From there, Pee Wee spent weekends in May at the Stuyvesant Casino as one of Jimmy McPartland's Stompers. In June, he was playing with Bobby Hackett, Sullivan and Wettling at Child's Paramount. Bandleader-drummer Ray McKinley hired Pee Wee to play a summer engagement on Coney Island, where jazz writer George Simon caught the band. Simon asked McKinley to play a guest spot on "Saturday Night Dancing Party," NBC's summer replacement television show. "There was no money involved," said McKinley. "Pee Wee played the blues on the show in the low register and playing with his clarinet bell right up against the bell of my cymbal. It made a weird sound, but it was good. It went over big."

An unusual opportunity occurred in Boston on August 13, 1952, when Pee Wee took part in a benefit for Storyville's musicians with a young Brandeis University professor, Leonard Bernstein. "That was one of the great nights," George Wein remembered.

> We had a fire and we lost everything so we decided to have a benefit. We called Leonard Bernstein, who we had done something with, and Lennie came in to play with us. He drove down in the rain from Tanglewood. Pee Wee came up from New York.

As *Down Beat* reported the event:

> Playing jazz piano for the first time in seven years, the young professor broke it up in a quartet performance with Pee Wee Russell, [drummer] Marquis Foster and bassist Jimmy Woode. Bernstein played a vigorous set in which he was also joined by J. C. Higginbotham. He scored a third time as soloist in a piano-thundering version of the "Honky Tonk Train Blues."
>
> Over 450 enthusiasts, overflowing into the lobby, jammed the club in response to an appeal to raise money for the Storyville musicians who were burned out of instruments, clothes and money in a fire that destroyed their summer residence on the North Shore.
>
> In addition to Bernstein, Pee Wee and Higginbotham, the Boston benefit was sparked by Ruby Braff, [clarinetist] Al Drootin, Joe Cochrane, George Wein and the entire Sabby Lewis band which appeared between sets of their regular gig at Sugar Hill.
>
> Musical high point earlier was the Bernstein–Pee Wee Russell dialogue, an event which excited Bernstein to plan a renewed interest in jazz activity. Even the usually reticent Pee Wee avowed it was a moving experience.

Wein recalled that the duet was based on the blues,

> but it wasn't really the blues. It was more like free jazz. Lennie kept playing and Pee Wee kept listening and following and fitting

in. Lennie didn't resolve anything. Pee Wee had to wait to see what he was going to do and fit himself in with it. Pee Wee had an incredible ear.

Returning to New York at the end of August, Pee Wee performed weekends in the jams at Stuyvesant Casino and with the Jimmy McPartland band at Lou Terrasi's. A photograph published in *Down Beat* shows him in the company of McPartland and trombonist Dickie Wells, demonstrating how to make the clarinet sound more like a trumpet with a cup mute.

With the encouragement of George Wein and Ruby Braff, Pee Wee began to make Boston a second musical home. In October, 1952, he was back, playing at Mahogany Hall, Wein's traditional club in the basement of the Copley Square Hotel (Storyville, with a more modern bent, was upstairs) with Braff, Russell "Big Chief" Moore, trombone; Ivan Wainwright, piano, John Field, bass; and Marquis Foster, drums. Pee Wee played several weekends in December back in New York at the Stuyvesant Casino, returning to Boston in January, 1953, when he recorded for George Wein's latest business venture, Storyville Records.

The 10-inch LP was his first session since the 1947 Eddie Condon all-star session for Decca. For his date, Pee Wee used Doc Cheatham, trumpet, Vic Dickenson, trombone; George Wein, piano; John Field, bass, and Buzzy Drootin, drums. His long-lost songwriting partner, Al Bandini, contributed one of the tunes, "Gabriel Found His Horn," and provided a vocal on it as well. The album also included one of Pee Wee's own compositions, "Missy." Tight arrangements by Dick Cary and the selection of the musicians and tunes combined to produce a new setting for Pee Wee, a sound that critic Stanley Dance would dub "mainstream" in a few years. Gone were the dixieland clichés of the years at Nick's and Condon's. There were subtle differences in Pee Wee's style as well. In the 1930s and 40s, he often played with such a fierce intensity that he sounded possessed. Here, his playing was rejuvenated but relaxed, the introspective reflections of a man finally at peace with himself. Growls and other vocalization techniques were used sparingly. His improvisations were just as unpredictable as ever but more lyrical. "His playing didn't have the brightness of his early records," said George Wein, "but he had another thing going."

Pee Wee's sound, however, still did not appeal to everyone. Listening to "Gabriel Found His Horn" in a blindfold test, bassist Oscar Pettiford said, "I wonder what was the matter with the clarinet player. He either sounded awful nervous or else he had some frogs in his mouthpiece or something." Cary's up-to-date arrangements provided the perfect setting for Pee Wee, but his reputation as the grand old man of the clarinet—a living legend—made it necessary for him to continue to appear primarily in dixieland settings. He was featured often at Condon's and frequently was seen as a guest star with various local bands in towns and cities on the eastern seaboard and in Canada.

Pee Wee was settling into domestic life in a way he never had before. Since his brush with death, both he and Mary seemed to have matured, and in the time since their reconciliation, their relationship had matured as well. With their marriage repaired, Mary and Pee Wee moved into a new apartment at 37 King Street in the Village, with their new dog, a schnauzer named "Winky." Pee Wee took long walks through the Village with Winky. The Village still was a small, relatively self-contained community, but one no longer mainly inhabited by artists. One was likely to run into people from all strata of society. On one of their jaunts, Pee Wee and Winky met Eleanor Roosevelt, who, it turned out, maintained a pied-à-terre in the Village and frequently walked her dog in the neighborhood.

Under Mary's care, Pee Wee continued to gain weight and confidence. He was heavier than at any other time in his life, almost 150 pounds. Although the doctors' strictures about total abstinence were ignored, with Mary's help he was able to moderate his use of alcohol, sticking primarily to a couple of glasses of ale a day and, on occasion, a brandy before bed. He was not nearly as reticent and withdrawn as he had been before his breakdown. He frequently expressed strong opinions on many topics, including jazz and his place in its history.

Pee Wee settled into a routine at the Stuyvesant Casino, playing there on weekends from January through April, 1953. The talent pool in New York was still rich, and participants in the free-for-alls at the Casino included most of the old timers and some of the young turks, such as Steve Lacy and Roswell Rudd, both of whom were still playing traditional jazz. The Casino, a cavernous catering hall virtually without ornamentation, was rented out for wedding receptions, bar mitzvahs and other celebrations during the week and was transformed into a jazz hall on Friday and Saturday nights. The clientele consisted largely of college students, many of whom came to dance. Those who did not dance either talked loudly at large tables, shouted to each other from the bar (which ran along one wall), or milled about in the aisles. The atmosphere was a far cry from Nick's in the thirties.

In February, 1953, Pee Wee was the guest star with Little Bobby Conway's Dixie Sextet at the Brown Derby in Washington, D.C. The band's regular clarinetist, Jimmy Hamilton (not related to Duke Ellington's clarinetist of the same name) was a devoted admirer of Pee Wee's clarinet style, but he did not play the gig—even if his clarinet did. Noticing that Pee Wee's clarinet was in even worse repair than usual, Hamilton lent him his own instrument. Even though Pee Wee played Hamilton's clarinet every night, he continued to bring his own "just in case." Hamilton said:

> Mary came to Washington with him, but she had the flu and Pee Wee wasn't eating right as a result. He needed her to take care of him.
>
> Just before we left the job the last night of the gig, Pee Wee remembered it was his wedding anniversary. So he gets two bottles of champagne, a bottle in each hip pocket, and he walked

out of the Derby. We get to the hotel and he invites me in to meet Mary and have a drink. Mary was in her negligee. First thing Pee Wee wanted to do was order ice and cigarettes. There were cigarettes all over the room. He said, "Mary, get Jimmy some ice so we can have a drink."

So we are sitting there, drinking warm champagne. Finally the ice comes and Pee Wee's happy: now we can have a real drink instead of warm champagne. But meanwhile Mary was peeved about the domestic champagne. She said to Pee Wee, very sarcastically, "Want to play the Anniversary Waltz?"

Mary was a very kind hearted, warm person, but sharp. She needled him all the time and it kept him going. They were always talking. They had learned that their marriage would last as long as they kept talking. She stimulated him. She would get his goat and he would counter-punch. I guess if you walked in on it, you'd say "Geez, these people hate each other!" But they didn't. They got along beautifully. It was a wonderful relationship.

So Mary starts in on Christmas of 1952, and tells me Pee Wee had to hock his horn for $15 to buy her a Christmas present. From the pawn shop he goes to the liquor store and buys $13 worth of whiskey. Then he has to borrow twenty-five cents from Mary to buy her stockings for $2.25 as a present. The stockings are the wrong size as usual. She says that Pee Wee always buys stockings and they are always the wrong size.

Mary knew Benny Goodman quite well and said that Benny told her that Pee Wee was the only clarinet player he was afraid of. Then she started laughing. She said, "Do you know, Benny Goodman plays classical, swing, back room, does all of that. And all that bum can do," she said and pointed at Pee Wee who was laying on the bed, "is play back room. You know, Benny could go into a back room and out-blow fifty Pee Wees. He could play Pee Wee out the back door."

Pee Wee's still laying on the bed. He says in a very low guttural mutter, "I'd like to get Benny in a back room."

Mary ignored the comment and told Hamilton that when Pee Wee almost died in San Francisco, she got a call from Eddie

Condon who thought that—even though Pee Wee and Mary
were separated at the time—she might want to start thinking
about funeral arrangements.

"The whole town got the word," Mary told Hamilton, "and
they were all set to go on a big drunk when Pee Wee died." Phil
Napoleon's band was playing at Nick's at the time, and the
musicians put a photograph of Pee Wee on one of the pianos and
draped black crepe all over it. The press took a photograph of the
band solemnly gathered around Pee Wee's photo which was to
have been used in the obituary.

The question came up at Nick's about where to bury Pee
Wee, Mary said. She pointed out that Pee Wee was officially a
Presbyterian, but that he was not religious, and since Mary was
Jewish, they thought of burying him in a non-sectarian ceme-
tery. The musicians decided that he should be buried next to
Nick—so Nick would have a friend. Mary told Hamilton that
Nick had purchased an extra cemetery lot for Pee Wee.

Pee Wee, who had been listening to all of this from the bed,
interjected that Nick was an Italian. Mary looked at him and
said, "No, Pee Wee, he was an American." "Italian," Pee Wee
repeated. "American," Mary yelled.

"All right, chum," said Pee Wee. "American."

"And don't call him anything else but American," said Mary.
"Italian," said Pee Wee.

Mary looked at Pee Wee, laying on the bed, and said, "And
that bum didn't die. He disappointed everybody. Condon hasn't
talked to me since!"

Following the Brown Derby engagement, the Russells returned to their apartment in New York. In May Pee Wee was
featured with Charlie Fisk's band at the Cafe Rouge in the Hotel
Statler. Fisk was a trumpet player whose career was with NBC
studio bands. Aside from the other guest star, George Wettling,
the personnel of the group remains a mystery. For at least part of
the month, Pee Wee continued at the Stuyvesant Casino on
weekends. In June he joined George Wettling's band, with

Johnny Windhurst, trumpet; Ed Hubble, trombone and Charlie Queener, piano, at Jack Dempsey's Restaurant in New York. The band played there for most of the summer, Hubble being replaced by Ward Silloway toward the end of the engagement.

Pee Wee returned to Stuyvesant Casino for the fall season, playing there on weekends off and on over a period of twenty-one months from 1953 to 1955. There were also many other gigs during that time. In January, 1954, Pee Wee settled into Condon's for a month. A Max Kaminsky date for Jazztone Records featured Pee Wee in the company of Miff Mole and Joe Sullivan, his former compatriots at Nick's. Pee Wee's beautiful low-register introduction to "Lonesome Road" and his heartfelt blues choruses on "Stuyvesant Blues" proclaimed that he had regained all his powers.

The motion picture camera caught Pee Wee in full flight during a gig at Central Plaza, located on Sixth Avenue between Sixth and Seventh streets, a weekend jazz emporium similar to the Stuyvesant Casino. The short subject, "Jazz Dance," directed and produced by Roger Tilton, opened at the Paris Theatre in New York to enthusiastic reviews. The film captured the exuberance of the audience—primarily college students. The camera was focused on the fans' high-spirited jitterbugging (this was just before the emergence of rock) and on a few professional dancers thrown in as ringers. Good shots of Pee Wee and the other musicians also abound, although the music, when it can be heard above the din, only adds to the chaos.

In April, 1954, Pee Wee participated in an all-star memorial concert for Frankie Newton, the superb swing trumpet player. Newton had been a close friend of both Pee Wee and Mary. Mary had become acquainted with Newton when both supported left-wing political activities during the early forties. The concert, held at Basin Street in New York, also featured Buster Bailey, Charlie Shavers, Sonny Greer, Miff Mole, Buck Clayton, Pee Wee Erwin, pianist Billy Maxted, Bobby Hackett, Willie "the Lion" Smith, Pops Foster, Cozy Cole, Pete Brown and Flip Phillips. More than two hundred fans were unable to get into the packed club.

Pee Wee was beginning to be taken seriously by the music world. The spring, 1954, edition of *The Clarinet*, a publication for professional clarinet players with heavy emphasis on "serious" music, carried an article by Pee Wee titled "The Development of Jazz Clarinet Playing." Pee Wee wrote:

When I was a kid in Muskogee, Oklahoma, back in 1915, my folks wanted me to study music and tried me on the drums and piano, but they didn't fit. To me, the clarinet was the best instrument of all—and I still think the same. So, when the instrument was finally decided on, I began taking lessons from Charlie Merrill, a local teacher whom I liked a lot. Later, when the family moved back to St. Louis, I studied with Toni Sarli, who was the solo clarinet in the St. Louis Symphony. Toni has been living in Los Angeles for years now, making pictures . . .

The instrumentation of those early jazz bands was usually drums, piano, banjo—sometimes a string bass was added—and a clarinet. The clarinet had the melody to himself, the other instruments being rhythm, which naturally made the clarinet player the big shot in the ensemble. It wasn't long before the cornet came along and nosed the clarinet up a third in the scale to make room. You can imagine how clarinet players felt about that. Now we can all look back and see that it was a good thing.

What happened—which is easy to understand now—was that we got tired of sitting on that third above the cornet who played the melody, so we began to fool around. We'd embroider the melody, dance around it, give it a lick or two, and the first thing you knew, you'd be running all over the place. When the trombone came along, its berth was made up and waiting—the fifth of the chord below. There you had your triad well spaced in three registers—the trombone playing fifth, the cornet the tonic, and the clarinet the third.

From then on, things began to pop. All the melody instruments started playing around and improvisation flourished. Of course you had to look out and keep out of the other fellow's way.

If you came a little too close to the cornet, you had to move up a bit, and he had to stay down out of your way, too.

I don't consider anything over six pieces a jazz band. The minute an arranger begins to take over (which is necessary when there are a lot of instruments), jazz is out. That's why young players have a more difficult time becoming jazz players today. When they are on their way up in the big bands, they've got to be note readers—and the heart of jazz is improvisation.

George Wein contacted Pee Wee in June, 1954, about his most ambitious undertaking to date: a "festival" in Newport, Rhode Island. Wein had been selected by the backers, Newport socialites Elaine and Louis P. Lorillard, to organize the event. The Boston pianist-entrepreneur brought together the very finest jazz musicians representing a wide variety of styles, from traditional to the latest developments in "cool" or progressive jazz, and spread the concerts over two days. The setting—upper-class Newport—was audacious. Aristocratic dowagers were apprehensive of "undesirable elements" trashing their carefully rolled lawns.

On Saturday, July 17, 1954, two bands opened the first day's concert. One was the Eddie Condon house band with Wild Bill Davison, Lou McGarity, and Peanuts Hucko. Bobby Hackett led the other band, which included Pee Wee, Vic Dickenson, and featured vocals by Lee Wiley.

The concert, attended by 6,000 enthusiastic fans, was scheduled to start at 8 p.m., but finally got under way a little after nine, following a long-winded welcome by Newport mayor John J. Sullivan. Stan Kenton took over as master of ceremonies and introduced Condon's band. They kicked things off with a spirited version of "Muskrat Ramble," and were followed by Hackett's group. Lee Wiley sang some of the songs most closely associated with her: "If I Had You," "Soft Lights and Sweet Music," and "I've Got a Crush on You." The set ended with both groups jamming together on "Bugle Call Rag" and "Ole Miss." A 40-minute segment of the concert was broadcast on

ABC and a tape recording of the entire event was made for the
Library of Congress. It was an auspicious beginning. Critic
Whitney Balliett wrote that "Russell was eloquent and as forth-
right as he can be." Following the Condon group and the
Hackett-Russell sextet were the Modern Jazz Quartet, Dizzy
Gillespie's quintet, Lee Konitz's quartet, the Oscar Peterson
Trio, the Gerry Mulligan Quartet and Ella Fitzgerald. The long
evening ended with the "dixielanders" teaming up with the
"progressives" on "I Got Rhythm." The jam session included
Pee Wee and Stan Kenton on piano. "I'd like to see Stan playing
piano behind Pee Wee Russell," Nat Hentoff was quoted as
saying backstage in the August 14, 1954, issue of the *New
Yorker*. "Progressive Prototype behind Old Conservative. What
dramatic contrast! A capsule history of jazz." "You can't tell,"
Kenton said. "We might find something together." A photo of
Kenton and Pee Wee, deep in discussion, ran in *Down Beat,*
with the caption "Who's understanding whom?" Hentoff's remark was more prescient than even he may have believed at the
time, for in just a few years Pee Wee would begin to explore a
new repertoire.

Wein's career as an impresario would be closely intertwined
with Pee Wee's. Wein respected Pee Wee's creativity and style,
and featured the clarinetist at nearly every Newport Festival
from then on. "Pee Wee's great talent was his awareness of
intervals," Wein said.

> Bix was important to him; Bix used different intervals, too.
> . . . Pee Wee, for me, embodied the jazz musician. He was a
> totally free spirit, musically. People accepted him on his terms.
> He never met anybody else's terms. He just kept playing as well
> as he could play. He was listening to new things all the time and
> absorbing it all in his ear, but at the same time he was just
> finding his note. That's how he spent his life: finding his note.

In October, 1954, Wein called Pee Wee to Boston to play at
Mahogany Hall in Vic Dickenson's band. During the engagement, Wein recorded the group for his Storyville label, under

the title "Jazz at Storyville." In mid-November, Pee Wee participated in a memorial concert for the trumpet star, Oran "Hot Lips" Page, at the Stuyvesant Casino.

The May 6, 1955, issue of the *New Yorker* reported that Pee Wee appeared with "Bobby Hackett's band," which included the incredible personnel of Miles Davis, Buck Clayton, Miff Mole, Urbie Green, Coleman Hawkins, Ben Webster, Johnny Hodges, Joe Sullivan, George Wallington, Pops Foster, Kenny Clarke, Sonny Greer and Jimmy Rushing, all playing at a concert at the Stuyvesant Casino, and the following night at Werderman Hall, 144 East 16th Street. Davis, Vic Dickenson, Kai Winding, Pee Wee, Hawkins, Hodges, Don Elliott, Wallington, Foster, Charles Mingus, Zutty Singleton, Jo Jones and Jimmy Rushing took the stand the following week at the Stuyvesant Casino, and appeared on May 14, 1955, at the Grand Ballroom of the Paramount Hotel.

Undoubtedly the mixing of the ultra modernists with the traditionalists was an attempt to appeal to everyone. Jazz appreciation was at a low ebb during the early 1950s. Almost all of the glamorous New York jazz clubs that had flourished during the previous three decades were gone. For the most part, dixieland was relegated to weekend gigs at large, noisy beer halls like the Stuyvesant Casino, the Central Plaza, and a few others where the emphasis was on dancing and socializing by the young college set, where only the musicians' loudest blares could be heard over the noise. Moveover, many kids were listening to a new singer with a funny name, Elvis Presley, whose career and the ensuing rock and roll craze spelled the end of teenage interest in jazz. Seasoned jazz fans considered dixieland to be tired, dated, trite, corny. Musicians like Pee Wee, so closely identified with Nick's and Eddie Condon's, found themselves out of it.

One good job that Pee Wee could count on each year was the Newport Jazz Festival. The 1955 edition had expanded to three days in July, and had Pee Wee playing with the Newport All Stars. The concert was recorded for broadcast by the Voice of

America. During the rest of that summer, weekend jobs in the Catskills and other resort areas at least provided frequent escapes from the city's heat. Before the fall, Pee Wee replaced Ed Hall, Eddie Condon's regular clarinetist at the time, who took off for a month's vacation. Hall's departure became permanent when, in September, 1955, he replaced Barney Bigard with Louis Armstrong's All Stars at the Crescendo Club in Hollywood. Pee Wee stayed at Condon's for a year.

In November, at the invitation of drummer Rollo Laylan, the Condon gang—Pee Wee, Wild Bill Davison, Lou McGarity, Gene Schroeder, Walter Page and George Wettling—found themselves in Miami, Florida, the land of the rhumba and the businessman's bounce. Laylan had played at Condon's in the mid 1940s, then headed south with New Orleans clarinetist Tony Parenti, settled in Miami and created a new persona for himself as "Preacher Rollo." Rollo's religion was jazz, and the nickname stuck because of his fondness for preaching about the glories of dixieland. He promoted a concert at Dade County Auditorium for the Condonites, with his band, The Saints, as the opening act. After the concert, both bands drove across Biscayne Bay to Miami Beach, where the Saints had a longstanding gig at the Shoremede Hotel. That night and the next, the local band gave way for the Condonites. Wild Bill Davison and Pee Wee were in great shape, but McGarity, who had been drinking heavily, was far from his best. Nevertheless, the event would be fondly remembered for many years by South Florida jazz fans.

Back in New York, there was a flurry of recording activity for Pee Wee at the beginning of 1956. In February, he cut four titles in a trio setting with Gene Schroeder and George Wettling. A few days later, George Avakian recorded the Condon band for Columbia. The resulting record was titled "Eddie Condon's Treasury of Jazz," a promotion for Condon's new book of the same name. The chapter on Pee Wee in the book was written by critic George Frazier. On the record, Pee Wee was featured on one of his own blues compositions, "Three-Two-One Blues." *Down Beat* reviewer Nat Hentoff gave the album three stars and noted that "the most creative soloist is Pee Wee and I hope George Avakian or Neshui Ertegun will give this man an LP to stretch out in."

Pee Wee sat in with Bobby Hackett's band at Child's Paramount for a couple of weekends in the spring. In April, 1956, the Condon band backed a long-lost Chicagoan, alto saxophonist Boyce Brown, on another album (Brown had entered a monastery years before and had taken the name Brother Matthew). By May, Pee Wee had wearied of the routine at Condon's and turned his job over to a young protégé of Sidney Bechet, Bob Wilber. Pee Wee rejoined Bobby Hackett's band at Child's Paramount that month and stayed through November, with the month of July off.

Just when Pee Wee's career seemed to be constricting along with the declining interest in dixieland, George Wein stepped in and gave it a whole new direction. Since his recovery from his illness, Pee Wee had been typecast as a "living legend," trotted out from time to time to commemorate the glories of America's musical past, then had been returned to his dusty display case and forgotten by all but a handful of fans. Wein knew that Pee Wee would shine in a more modern setting. Although he was not the producer of the concert at Music Inn in Lenox, Massachusetts (Nat Hentoff and Bill Coss were the producers), Wein had the inspired idea of contrasting the old with the new by having Pee Wee play a blues duet with the progressive West Coast clarinetist Jimmy Giuffre. The event took place on August 30, 1956. Wein recalled:

> I had several reasons for this other than my desire to hear these two wonderful jazz musicians play together. For years, Pee Wee has been one of my favorites. Some of my most exciting musical experiences have occurred while working with this native American from Muskogee. It has always annoyed me that so few others recognize or even bother to listen to the genius that is Pee

Wee's. I've always felt that even the people that are Pee Wee
fans don't really understand his music. Most of them think he is
cute, amusing or fun, etc. Having worked with traditional Chicago groups most of his musical life, his talent was usually buried
in the ensemble.

At the Music Barn, the opportunity presented itself for Pee
Wee to be heard under the best possible conditions. For over
eight minutes Jimmy and Pee Wee tossed the blues back and
forth to each other in bursts of musical emotion that I hoped
would never end. The success of this session only highlighted
another of my reasons for being concerned with this memorable
encounter. Both these men played the blues. Although one was a
comparative youngster, a product of the last nine or ten years of
jazz, and the other grew out of the Bix Beiderbecke Chicago era,
their supposedly conflicting styles didn't serve as a detriment to
the general message of the music.

There was one tinge of regret. While all the musicians
seemed to think the whole idea was fun, none seemed to be
concerned with absorbing anything from Pee Wee's playing.
Their remarks ranged from "he has the old jazz feeling in his
playing" to "man, he sure can play the blues," to "his tone
bothers me, but I enjoyed the old man."

I heard no mention of his amazing conception of time, which
shows a knowledge of the whereabouts of the beat equaled by few
musicians. And nobody enthused about his wonderful choice of
notes. It would be interesting to see an analysis of the notes used
by both musicians in their explorations . . . I wonder just how
much more harmonically advanced are the modern sounds of
Jimmy's beautiful clarinet to the "old" sounds of Pee Wee.

Selections from the concert were issued on an Atlantic LP. It
was the duet (the clarinetists received understanding support
from Wein, piano; Oscar Pettiford, bass; and Connie Kay,
drums) that captured the interest of the jazz press. Wrote the
reviewer of the LP in *Down Beat:*

> The high point is an eleven-minute Pee Wee Russell–Jimmy
> Giuffre duet. Like a pair of spiders spinning interlocking webs,

the two clarinetists develop a set of fragile blues arabesques that are at once delightful and instructive.

Russell, ever the non-conformist, displays his genius for building oblique choruses of arresting beauty, finding the unlikely right notes just before it is too late to play them at all. Giuffre speaks even more softly than Russell but unlike Pee Wee, does not carry a big stick . . . This night belonged to Pee Wee Russell.

In a lesser performance, the entire endeavor would have been dismissed as nothing more than an interesting novelty, but Pee Wee's intense playing demanded to be taken seriously. The record opened the ears of many critics and jazz fans who had consigned him to the past. Professionally, the performance freed Pee Wee at last from the confines of Condonism and opened new vistas for him. The new settings rejuvenated his creative powers as well. No longer hammered like a round peg into square holes labeled "Chicagoan," "Nicksieland," "Dixieland" and "Condonstyle," Pee Wee was now—in Duke Ellington's phrase—simply "beyond category."

Pee Wee had more than just recovered from his illness five years before. He had been reborn physically, mentally and musically. Although he would never be robust—he would have relapses, especially when, on occasion, he lost control over his drinking—his condition was certainly much improved. He was more at ease with himself, felt secure and happy in his life with Mary and was being taken seriously by the fans and critics. With Mary's—and Wein's—help and guidance, he was entering the most successful period of his life.

11 • • • •

New Fame

*J*azz is a young man's art. Few great jazzmen have managed to maintain the peak of their playing much past their mid-thirties. Yet, at the age of fifty and for the next decade, Pee Wee's playing blossomed. His improvisations, although lacking some of the fierceness they once possessed, were as original and creative as ever. English critic Charles Fox wrote: "It is Pee Wee's ability to communicate the sheer zest of musical creation that makes him sound so youthful."

Pee Wee was as busy as he wanted to be with recording sessions, jazz festivals and night-club bookings. He played a wider variety of musical styles with a wider variety of musicians than probably any other jazz musician. At a time when dixieland jazz had lost much of its popularity, Pee Wee had ceased to be considered merely a dixielander—if, indeed, he had ever been one. Of course, most of the jobs still revolved around dixieland ensembles, but on occasion even those gigs could be different. In January, 1957, he appeared with a George Wettling group (Max Kaminsky, Ray Diehl, trombone, Dick Cary, piano) that played on the same bill with the Savannah Symphony Orchestra. One of the local newspapers reported that the Symphony "held its own auspiciously last night." Pee Wee's next gig later in January

214

returned him to Condon's Greenwich Village hangout for what would be the last time. The glib guitarist was absent from the premises, having taken his regular house band on a tour of England. The band filling in at the club was led at first by Jimmy McPartland, but he was replaced by Johnny Windhurst.

In January, Ruby Braff and Pee Wee cut an "experimental" session for George Avakian, who was producing the album for the Epic label. The repertoire included some modern titles and arrangements. According to Avakian, Coleman Hawkins and Vic Dickenson were also among the personnel. Avakian, Pee Wee and Braff remembered the date as a great success. Pee Wee wrote to Jeff Atterton, an English fan adopted by Pee Wee and Mary as a pen-pal:

> It's the first time in my life that I'm excited about records. "We blew our brains out . . . Ruby is sailing about those records. He got us out of bed a few days ago at two in the morning so he could talk about them. He wanted to try some new stuff, and George Avakian let him do whatever he liked. It was going to be either the world's lousiest or the best. Ruby had been under a strain, so we relaxed with a beer and coffee and talked.

The record, however, was never released. Apparently, the tapes were destroyed by another Columbia producer in a corporate battle.

On March 3, Pee Wee and Windhurst played a concert at the State College of New York at New Paltz. Neil Timberlin, a student at the college, sat in on bass. Recording sessions in March and April, 1957, contrasted old settings with new. In alternating sessions Pee Wee recorded with a Bud Freeman group re-creating "Chicago" classics and with a mainstream Ruby Braff group playing tunes associated with swing trumpet star Bunny Berigan. While Pee Wee's playing sounds routine— even somewhat bored—with Freeman's band, he dug in on the swing era titles. Braff, one of Pee Wee's "kiddies" in his 1952 Storyville band, had matured into a spectacular cornet player in

the intervening five years and had grown close to Pee Wee both
socially and professionally. On July 5 the two musicians appeared at the Newport Festival with the rhythm section from
their working band: Nat Pierce, piano; Steve Jordan, guitar;
Walter Page, bass; and Buzzy Drootin, drums. Braff introduced
the set by informing the audience they would hear "no psycho-
logical or psychotic music . . . no fugues," in reference to the
music of the progressive players that had preceded them. The
critics responded with praise. Jazz critic Dick Gehman wrote:

> The man they all said could never make it again made it again
> yesterday afternoon and in the process broke up everyone present
> at the fourth annual Newport Jazz Festival. They made everyone
> in the audience sit up and look in wonderment. First at each
> other and then back to the stand. Everybody in the band was
> blowing his best, but Pee Wee's best was better than anyone
> could believe.
>
> Bending over his horn, his face contorted, his neck muscles
> ready to burst, he blew as though he had just discovered that he
> could. And that, after all, is the essence of jazz: a spontaneous
> and original air of discovery . . . Everyone present agreed the
> day belonged to Pee Wee. The question on everyone's lips was:
> Why isn't this man, one of the all-time jazz greats, given the
> recognition he and his music deserve?

Whitney Balliett, writing in *The New Yorker,* noted that

> In "Nobody Else But You" Russell took a gently, barely audible
> low-register solo that was the equal of any improvisation during
> the whole weekend.

Writer-photographer Burt Goldblatt echoed the assessment:

> Pee Wee, looking frail and sensitive, played with a completely
> blues-rooted sound. The feeling that he projected when he soloed
> on "Nobody Else But You" playing in a low register, had a fresh-
> ness and directness that touched the audience deeply.

During the year, Pee Wee recorded a solo album, supported by a
rhythm section consisting of Pierce, Jordan, Page and George
Wettling, for a small independent label. Two very different

takes of "I'd Climb the Highest Mountain" were included, both magnificent performances. "It reminds me of Bix," Pee Wee wrote to Jeff Atterton about the song.

> Of course I like the tune anyway, but I always have a picture of Bix fooling around with it on the piano. That's when we shared a house. Or was it an apartment? Anyway it was a dump of some kind. Sometimes I would pick up my horn and join him. We must have done that on a lot of tunes, but it's "Mountain" that stayed in my mind.

Pee Wee also included in the album two of his own compositions, "Muskogee Blues" and "Pee Wee's Song" (the latter issued as "Pee Wee's Tune" in subsequent versions).

In August, Ruby and Pee Wee opened for two weeks at the Village Vanguard with Pierce, Jordan, Page and Bobby Donaldson, drums. The *Down Beat* reviewer who caught opening night reported that

> Russell, a persuasive statesman on the clarinet, was in excellent form. His solos were soft edged, thoughtful, often humorous, and always urbane. [The audience reaction was] very warm, particularly for Pee Wee, whose solos terminated with appreciative applause . . . Considering the quality of the music and the genuine happiness created, there is more than a demand for this group. There is a definite need.

At the conclusion of the Vanguard gig, they opened Jazz City, a new restaurant-bar just off Times Square. The band was supposed to appear there for a month, but the job was cut short by two weeks. "I'm not too discouraged, and I hope Ruby isn't," Pee Wee wrote to Atterton.

> It was a lousy job. The bosses were old time semi-hoodlums, prohibition-days style. With those characters the word musician automatically goes with jerk. At any rate, Ruby wanted out, and they were just as glad to see us go. I guess they didn't expect jerk musicians to get just as tough with them. One time Nat Pierce said to one of the hoods: "Go ahead, hit me. I'll call my lawyer and next week I'll own the joint." I admit I almost felt sorry for

the inch-brow. He looked helpless and frustrated and of course
didn't do a thing.

Nat Pierce remembered how the incident began.

We hadn't gotten paid for the first week and a half, so Erroll
Garner comes in one day, and the owners, godfather types, were
trying to woo him to come in there. After talking to them for a
while, he came over to our table and asked us how we were
doing. I told him fine except they were not paying off on time.
With that, he jumps up, says: "Excuse me, I got to go to the
men's room," runs out into the street, jumps in a taxi and is
gone. The next thing I know this big gangster yells at me, "Hey,
piano player," and hauls me into the back room where he throws
a foot-long cigar at me. It was like a torpedo aimed at my head,
but I ducked. Later, we did get paid. Those guys didn't fool
around with Pee Wee. They took care of him.

Around this time, Eddie Condon's Greenwich Village Club
closed its doors when New York University decided to build a
library on the site. Condon moved uptown into a club formerly
known as Bourbon Street in the Sutton Hotel, 330 East 56th
Street. Pee Wee played there occasionally through the years, but
his career was headed in a different direction. On December 1,
for example, he was featured with the modern guitarist, Charlie
Byrd, at the Bayou Restaurant in Washington, D.C. The afternoon concert was sponsored by the Washington Jazz Club. Pee
Wee was his usual nervous self when he took the guitarist aside
before the concert and asked: "You remember that old Fats
Waller song, 'My Fate Is in Your Hands?' Well, that's you
today." The concert was a smashing success.

Four days later, Pee Wee was in Columbia's studios, rehearsing for another triumph, the highly acclaimed Sound of
Jazz television show. Hosted by television critic John Crosby,
the show brought together many of the biggest names in jazz
history, forming a time capsule of the state of the art in 1957,
from the New Orleans master trumpeter, Henry "Red" Allen, to
the avant-garde sounds of Thelonious Monk, and including

among the many luminaries Billie Holiday, Lester Young, Coleman Hawkins, Roy Eldridge and Count Basie. The musical advisers were Whitney Balliett and Nat Hentoff.

Hentoff remembered the electrifying duet played by Pee Wee and Jimmy Giuffre at the Music Inn concert more than a year before and suggested they repeat the performance on the show. Pee Wee appeared first on two numbers with Red Allen's group—a far from traditional New Orleans ensemble—which included the explosive trumpet and vocals of the leader, with Rex Stewart, cornet; Vic Dickenson, trombone; Coleman Hawkins, tenor sax; Nat Pierce, piano; Danny Barker, guitar; Milt Hinton, bass; and Jo Jones, drums. The sensational hour closed with the Giuffre-Russell blues duet, accompanied by Barker, Hinton and Jones. Unfortunately, the program ran overtime, and the performance had to be faded out. (Recordings made at the rehearsal three days before included a complete performance, which was issued by Columbia Records.) The network television performance introduced the "new" Pee Wee to a larger audience and proved that his previous collaboration with Giuffre had not been a fluke. Pee Wee said of Giuffre:

> I like to hear him play. I don't know what it is, but his playing is as elusive as his personality. There's something interesting about him. You can talk to him and it's almost like he wasn't there. I sure had a ball playing with him on the show. We got along very well together.

George Wein summoned Pee Wee to appear at Mahogany Hall in the Copley Square Hotel for the Christmas holidays. Gerry Mulligan's quartet, with Bill Crow on bass, was playing upstairs at Storyville. Mulligan suggested that he write an arrangement on "Auld Lang Syne" which both bands could play on New Year's Eve, and Wein thought the idea was excellent. The two bands rehearsed the arrangement during the afternoon of New Year's Eve. "Pee Wee made a lot of suffering noises because he was worried about having to read music," Crow said.

He sounded fine at rehearsal, but he continued to worry. That night, both bands jammed on a few tunes, and as the hour of midnight approached, Mulligan signaled the musicians to get out his chart. Pee Wee could not find his part. We searched everywhere, Crow recalled. With midnight only seconds away, the clarinet part was still missing, so we just faked "Auld Lang Syne." Gerry was disappointed, but the audience, unaware of the arrangement we hadn't played, was content.

As we left the stand after the set, I passed the chair where Pee Wee had been sitting. There lay the missing part. The crafty bastard had been sitting on it all the time.

Pee Wee kicked off 1958 with a commercially oriented recording session led by Jimmy McPartland, in which Pee Wee soloed on Dick Cary's big band arrangements of "Gary, Indiana," and "The Wells Fargo Wagon" from the then-current Broadway musical hit, *The Music Man*. On January 27, 1958, Pee Wee and Wild Bill Davison were part of a jazz band that was flown to Omaha to play a private party for Gilly Swanson of Swanson's Frozen Foods. The company had just been sold to Campbell's Soup. On February 1, Pee Wee appeared at an all-star dixieland festival at Carnegie Hall, reminiscent of the huge Condon bashes. The headliner was comedienne Dody Goodman and the event was billed as "Dody in Dixieland." Pee Wee explained to Jeff Atterton:

> She's the dumb blond bit on the Jack Paar show. I doubt if you know who Jack Paar is. And I'll be damned if I know what she'll be doing with us, but it's a job and I'll get paid for it, I hope. Dody Goodman is my pet peeve on television. I've been forced to watch her when Mary wants to see Hermione Gingold, who makes that show occasionally, and I don't get her at all. Maybe I won't have to be too close to her.

Portions of the melee appeared on an album.

Around mid-Frebruary, Pee Wee recorded an important album under his own name, Nat Pierce supplied the arrangements. The album was appropriately titled "Portrait of Pee

Wee," and had a loose, swinging sound, with Pee Wee in his most lyrical form. One of the tunes was the first recording of "Pee Wee's Blues," a hauntingly evocative composition that would become a much requested part of his repertoire from then on. "The song came about in the studio," recalled Nat Pierce, who received co-composer credit on the piece. "He played this little figure and I kind of organized it a little bit and that was it. We didn't think we were creating a masterpiece or anything." Nevertheless, the beautiful song has had a life of its own, being featured by many other clarinetists through the years. Critic Martin Williams awarded the album four stars in *Down Beat*.

Ruby and Pee Wee had been rehearsing a slightly different group which included Bob Wilber on tenor, Mickey Crane (Carrano) on piano, Bill Takas on bass, and Buzzy Drootin, drums. The band had hopes for a European tour, but it apparently never performed in public, although Pee Wee and Wilber appeared together in May with Max Kaminsky for a gig in the Gothic Room of the Hotel Duane. Pee Wee would return there in July with Braff and an entirely different band.

Braff and Pee Wee flew to Toronto in May to appear on a Canadian television program, Cross-Canada Hit Parade, where they were featured with local musicians in striped shirts, suspenders, and straw hats. Nevertheless, Braff and Pee Wee gave "Oh, Lady Be Good" a good workout and accompanied singer Phyllis Marshall on "Who's Sorry Now?" At that point in Pee Wee's career, however, straw hats appeared even more anachronistic than February's Carnegie Hall chaos. After the videotaping, John Norris, publisher of the Canadian jazz magazine *Coda,* took them to a "more congenial atmosphere" with other musicians, where they wound up participating in an all-night jam session that was, according to Norris, "miraculous."

The next day, Pee Wee flew home to more television exposure. A New York jazz radio personality, Art Ford, began a series of televised "jazz parties" in May. Pee Wee was featured on the first one and was a frequent guest on subsequent programs, paired with musicians like Rex Stewart, Charlie Shavers, Tyree Glenn, Lester Young and Coleman Hawkins. The ninety-minute programs utilized the still-emerging technology of stereo by having one channel broadcast on an FM station and the other channel heard on an AM station. The programs originated live on Thursday nights from the Newark studios of WNTA-TV and were seen only in the New York area. Reviewing the first telecast, critic Whitney Balliett, usually a strong supporter of Pee Wee's, wrote: "Russell, who is now in the queer position of being at once a legend and a neglected musician, sounded defensive and forced in 'Sugar,' done as a solo . . . ." Such criticism did not bother Pee Wee, who seemed more confident than ever. Though his marriage to Mary was still contentious at times, he usually complied with her "two quarts of Ballantine ale and a bedtime brandy" rule. Mary, still professing that she didn't like his playing, was his ferocious protector against the outside world.

Pee Wee was also putting on weight. Before his illness, he weighed around 125 pounds, but now he was topping 150, still not exactly stout for a six-footer, but hefty for Pee Wee.

> I'll tell you how I got fat [he wrote Atterton]. "I drink at least two quarts of beer every day. I never got drunk on beer in my life. I drink an egg nog every day. Two or three eggs, a pint of milk and sugar. And I have a glass of milk with every meal. I eat breakfast and dinner and I have a sandwich or crackers and cheese late at night when I'm watching TV. When I'm away from home I lose weight. I don't eat and forget to take my daily vitamin pill. And I drink too much whiskey. But when I'm home, I'm a healthy guy. I've gained most of my weight around my middle. But, what the hell—I'm a middle-aged guy and it's time I got fat. And I smoke two or three packs of cigarettes every day of my life. I'm as nervous as a bastard and if I didn't smoke I couldn't stand it.

A few months later he reported to Atterton:

I became a wagon kid about two weeks ago. Vitamin shots from the doctor and stuff like that. Beer too. No whiskey. Drinking isn't fun anymore. It's too much of a burden.

The image of Pee Wee as somewhat befuddled lived on in stories such as the following, which appeared in the jazz press in November, 1958:

Each day our little heart goes out to somebody and today it goes to Pee Wee Russell. He had to go to the union yesterday to pick up a check for a gig he played. He picked it up and was studying the figures, also wondering where he could cash it, when he stepped off the curb at Broadway and 50th Street. He was quickly handed a jaywalking ticket. Entering the subway, he started to figure out the jaywalking ticket and lit a cigarette to help himself figure. He had only two puffs before he got a ticket for smoking in the subway.

But the condescending view of Pee Wee was becoming as much a thing of the past as the old doubts about his playing. "Pee Wee was nervous because he was very distrustful of people," said Ruby Braff. "He didn't know if it was going to be something funny at his expense or what. He would appear nervous, but he was the perfect gentleman. He was also the perfect scapegoat and very often he would sort of play the part until he'd get mad and tell them to go screw themselves."

There was no question, by this time, of Pee Wee's musical status, and it was natural that he was included in an all-star group of Buck Clayton, Jack Teagarden and Lester Young at the 1958 Newport Jazz Festival. For a group of living legends, the band played a very feisty set, winding up with a stomping "Jump the Blues" that brought the crowd roaring to its feet. Even Young, who was very ill, sounded vibrant and the others gave exceptional performances. Everyone was impressed except Pee Wee.

It wasn't much of a set. [he wrote to Atterton] It began to rain and I was glad when it was over. We had three numbers to do.

We talked them over first. We would start with Royal Garden, go on to a medley which Lester was to start, me second, Buck and Jack would take it out and then some blues. We get up on the stand and no Buck. And we sure as hell can't go looking for him. So we started without him. You know about the breaks on Royal Garden. I began playing the melody. Buck showed up in the middle of the number. In the meantime Lester is looking out to the audience, having nothing to do with any of this. That may have been his way of showing he didn't like the company he was in, but it sure as hell wasn't professional.

We started the second number, the medley. Lester started it and forgot to stop. He went on and on with his tune and when he decided to finish, the set was about over and Jack signalled to take it out. On the third tune, the rains came and we said to hell with it all over the place. So that was Newport, and I've never liked anything less.

Wein recalled Young was "very upset because he didn't have his own band. He walked out on the stage and said, 'Here I go auditioning again.'"

Wein's next production was the French Lick (Indiana) Jazz Festival, where Pee Wee performed on August 2, 1959, with Jimmy McPartland's All Stars. The front line also included Buck Clayton, Bud Freeman, and Vic Dickenson. Jimmy Rushing joined the group for a few numbers and Jack Teagarden, who had played with his own group earlier in the evening, couldn't resist also joining in. On Rushing's "Harvard Blues," Pee Wee played two spellbinding blues choruses which were followed by two equally inspired from Teagarden. "We played our little hearts out," Pee Wee wrote to Atterton. "They kept us on for an hour and ten minutes and I enjoyed every minute of it."

Around this time, Peanuts Hucko sold Pee Wee his alto saxophone.

Almost new, and a beauty. [Pee Wee wrote to Atterton] Ruby insists that it is the best instrument in the world. Toots Mondello spent months picking out the right horn for Peanuts. I'm sold on it and can't keep my hands off it. I fool around with that axe all the time. Monday I will buy an exercise book because I've become serious about the alto. Of course the clarinet is my real love but I like that new kid.

Braff encouraged Pee Wee's alto playing. The cornetist lived in the same building, as did another jazzman, trumpeter Johnny Windhurst, and card games became a frequent ritual. "Pee Wee was a fine musician," Braff, a most exacting musician, said. "He played all the horns and could double on everything. One time I had him playing alto with me. It was some place in New Jersey—a strange place: there was nobody there. His alto playing was really beautiful. He was marvelous at it. He sounded just as different. But he couldn't be bothered with it so he gave it up."

A more traditional-sounding Pee Wee performed in the refreshing coolness of the Catskills on July 28 when Alex Grossman, an insurance agent there, brought Pee Wee to Paddy's Tavern at Leeds, New York, for a trio performance with Gene Schroeder, piano, and George Wettling, drums. The three-hour session was sponsored by the Alumni Association of the Catskills Boys' Club. A crowd of 150 fans heard the trio run through numbers like "Body and Soul" and even "When the Saints Go Marching In."

A more modern setting presented itself in August when Pee Wee recorded with clarinetist Tony Scott on a album dedicated to Fifty-second Street.

> Tony is a sweetheart of a guy. [Pee Wee wrote to Atterton] He was working at the Half Note, a fairly new club, and it's only a couple blocks from where we live. Mary and I walked over one night (Tony had sent a note asking me to drop in sometime), and we had a great time, and I walked out with a record date. It was supposed to be something representative about Fifty-second Street but somehow Tony and I got together on a couple of sides. I don't know that it had anything to do with Fifty-second Street,

but we both enjoyed what we did and it just might turn into something.

Later that month he opened the Randall's Island Jazz Festival with Max Kaminsky's band. In September, he appeared on another Art Ford jazz party, this time with Coleman Hawkins and Lester Young.

Pee Wee's 1959 Thanksgiving was spent playing in the Macy's parade in White Plains; Benny Goodman played the New York City parade that year. The following night, Pee Wee played in another huge extravaganza, an "Ivy League Bash" put together by Stan Rubin at the Hotel Roosevelt. Playing sets throughout the hotel were several college dixieland bands, Rubin's Tigertown Five, and groups featuring such New York musicians as Pee Wee, Rex Stewart, cornet; Buck Clayton, trumpet; Wilbur DeParis and Cutty Cutshall, trombone; Marty Napoleon, piano; and trumpeter Randy Brooks, who played the penny whistle. Stan Rubin "plays the clarinet," wrote Pee Wee to Atterton, "and I think that's why he's making me leader for one of the bands. I get leader money and he's loyal to his instrument. Good boy!"

Toward the end of the year, Pee Wee became a member of ASCAP.

> Around here it's a big thing, [Pee Wee wrote to Atterton] I've been fooling around with tunes for a long time but never had the nerve to think of myself in terms of song writing. Then Mary gave me a nudge when she pointed out that every time I take a chorus I compose a tune. And she also said I can't get arrested for trying. So I did some stuff. Even had the gall to put a couple of them on "Portrait of Pee Wee."
>
> It isn't easy getting into ASCAP. They are selective. As a matter of fact, it's the toughest thing in the world to get in. It took plenty of prodding but I did call a big guy I know who's a big wheel in ASCAP. It wasn't easy for me to ask. I hate asking favors. I suppose it has something to do with that I might be turned down and I wouldn't like myself for asking. But I was

pleasantly surprised. He was great. He told me that he would call me.

Wait. I didn't tell you that I called him months ago, before the summer. He told me that the membership board meets in November. I didn't call him again, thought he had forgotten all about me and that was that. A week ago yesterday he called me and told me to get my compositions in order. I won't go through all the details but I had to pass before three boards and yesterday I officially became a member. And it's a hell of an incentive. I'm working on things. I'm not dreaming about silly stuff like making the hit parade and making a million bucks but I do want to leave something when I die. I mean a tune that I might be remembered for. For chrissake, now I'm getting sentimental. That's awful. Let's forget the whole damned thing. I do want you to wish me luck.

December was spent in Boston at George Wein's Storyville. Pee Wee worked with Buck Clayton and Vic Dickenson. Bud Freeman joined them for the first week, and Gerry Mulligan came in with a band that played opposite them. Pee Wee was very impressed with Mulligan's trumpet player, Art Farmer, and said he would like to use him on a recording date. Mulligan enjoyed Pee Wee's playing and sat in with the band frequently. George Wein said he had never heard Pee Wee play better.

When Pee Wee returned home from the job, he was "in the dog house. I came home a day late and drunker than I should have been," he wrote to Atterton. "I'll tell you the reason. I haven't been able to tell Mary but this is one way of getting her to listen." (Mary typed all of Pee Wee's correspondence and frequently interjected her own comments.)

I waited for a plane and they were grounded. [Pee Wee continued] A seventy-five-mile-an-hour gale and not one plane took off. I waited in a saloon. And I don't drink ginger ale in a saloon. I spent too much money. I got too drunk. Dinner was spoiled. And all the rest of that jazz. Every now and then Mary reacts to things like a woman and this was one of the times. Just when I was as sick as a dog. But isn't that the way it always is? Anyway,

Mary's normal again. My kind of normal. She got me some ale this morning. And I don't mean ginger. (Even Winky didn't talk to me.)

Pee Wee's next project was to record some of his new compositions for Dot Records. Although writing the tunes did not present a problem ("These damn things come so easy I'm ashamed of myself," he said), he was panicked by the idea of recording them.

> He wrote most of them in about two weeks. [Mary wrote to Atterton] He has spent weeks and weeks with Dick Cary getting them in shape. I have never known him so unsure of himself, and for the very first time I'm sorry for him. He is over-anxious to have a good date. And he knows that nothing will go right. I wish it were next Tuesday so the whole thing will be over with and I can tell him that it doesn't matter . . . Pee Wee is a hard working angel. When he has something to do he never goofs. And he has been working on his Dot date so very hard. I haven't allowed myself to think that it will be a lousy date. I hate to have Pee Wee hurt. This date is so important to him so you have your fingers crossed, too.

Pee Wee recruited Buck Clayton, Vic Dickenson, Bud Freeman, Eddie Condon, Bill Takas and George Wettling for the date. Cary was on piano. Then, just before the date (February 23, 1959), a Dot executive, a Mr. Leslie, called Pee Wee in Cary's apartment to tell him the date was postponed.

> Leslie told Pee Wee that Bob Thiele wants to supervise the date personally and because he was flying to the west coast on Monday the date will be cancelled. [Mary wrote to Atterton] "Not entirely cancelled, but pushed up to some time in the future.
>
> So Pee Wee, who can be louder and more ferocious than anyone I know when he is crossed, exploded. He bawled that Thiele can go to the Coast on Thursday. That he's worked like a dog for six weeks writing songs and getting them in to shape. That he's got the men he wants now and he may not be able to get

them again. And to hell with it all over the place. He'll take this
date someplace else and they'll be glad to have it.

Mr. Leslie was, to say the least, startled. Small time musicians don't send Dot to the devil. And Pee Wee *has* a reputation
for mildness. He told Pee Wee he'll call him back in five minutes
and Pee Wee graciously allowed him ten. And it was in. Thiele
did push his trip up to Thursday. Thiele did supervise the date
himself. Thiele did send out for a case of ale. Thiele did fall in
love with Pee Wee's tunes.

When I heard Pee Wee's key in the lock yesterday I knew I
shouldn't say a word to him. He came in, Winky greeted him
wildly, and I said to myself—he's tired but happy. I gave him
time to get his coat off, got him some cold ale and then inquired
casually—How did it go? (I was making like the perfect wife.)
And Pee Wee's reply—"it was a bitch"—reassured me. That's
Pee Wee's highest praise. And then he told me all about it. How
kind Thiele was and how satisfied with the date Thiele is.

A few days later, Pee Wee was back in the studios, and this
time no special preparations were required. George Avakian was
producing (for Warner Brothers) an album to commemorate the
twentieth aniversary of the first album he had produced: the
Chicago Jazz set for Decca (which, in turn, commemorated the
McKenzie-Condon Okeh recordings of the 1920s). Pee Wee ran
through the familiar repertoire once again.

On March 16, 1959, Pee Wee was with Dick Cary's band
when it played the Washington, D.C., Jazz Jubilee, a benefit
arranged by a group of government officials' wives, including
Mrs. Dwight Eisenhower, Mrs. Richard Nixon and Mrs. Earl
Warren. The event was narrated by Willis Conover. More than
1,500 cheering fans attended, raising more than $10,000 for the
charity, Friendship House. Two other bands, a New Orleans
group led by drummer Paul Barbarin and a modern big band
directed by Toshiko Akiyoshi, also participated. The proceed-
ings were recorded by Mercury but never issued.

For the first week of April, Pee Wee appeared in Toronto

with Mike White's Imperial Jazz Band at Basin Street in the Westover Hotel. White had been a protégé of Willie "the Lion" Smith and came highly recommended by the great stride pianist. Patrick Scott wrote in the *Toronto Globe and Mail:*

> As a Pee Wee Russell fan of long and ardent standing, and more than ever as a result of his memorable contributions at the Westover this week—I am prepared to admit that he probably is an acquired taste. For anyone reared on the liquid sounds of a Bigard or a Goodman, or even on the earthier approach of an Edmond Hall, an initial encounter with Mr. Russell must come as something of a shock; the sounds he produces are like nothing else in music let alone jazz—but to me, in this age of high-gloss, production line jazz, they are like a breath of fresh, clean life-giving air.
>
> Mr. Russell's tone—which is what has always seemed to bother his non-fans most and which has been described at one time or another as croaking, creaking, sour, tortured, inimitable and incredible—is undoubtedly one of the most maligned in jazz. In reality, and particularly in the lower register, it is a thing of beauty—alternately husky, wispy, full, thin, clean and dirty, but never, for a minute, empty.
>
> His technique, which has been likened to that of a 10-year-old boy, is more than adequate, actually, for Mr. Russell's purpose, which is not to ripple off glittering choruses from memory and long rehearsal but to feel his way along as he goes: if he fumbles occasionally, it is not from technical incompetence but because he has changed his mind in midstream, as good jazzmen often do.
>
> Which brings us to Mr. Russell's conception, his greatest single attribute. In originality, imagination, lyricism and daring, Mr. Russell is in a class, these days, by himself. To even attempt to anticipate which way one of his phrases will turn next is one of the more suspenseful games in jazz; indeed, even Mr. Russell sometimes seems startled by the outcome.
>
> If you are fortunate enough to have heard him this week, preferably threading his low register way through something on

the order of Ain't Misbehavin' or Black and Blue, you have heard one of the most creative men in jazz.

While in Toronto, Pee Wee also appeared on a CBC television talk show Tabloid, where he was somewhat condescendingly interviewed by Joyce Davison. Delving in Pee Wee's childhood, Miss Davison asked, "Were you a bad boy, Pee Wee?" "Oh, yes," he replied, squirming under the hot television lights, "very bad." Pee Wee's childhood surfaced again in his next gig, for which he journeyed to his birth place, St. Louis, with Bud Freeman's group—Buck Clayton and Vic Dickenson plus a local rhythm section. For ten days beginning April 27, 1959, the band did turn-away business at the Boulevard Room in the Hotel Sheraton-Jefferson. He was treated as a returning hero by the local press. A Mrs. Gould came by and told Pee Wee she had been a friend of his mother's and had held him in her arms when he was two hours old. Red McKenzie's sister came in to reminisce about the old days in St. Louis. Frank Orchard and other musicians in the area paid their respects. One disconcerting visitor, however, was his old flame, Lola, who had returned to St. Louis after their affair broke up. Her blazing red hair had turned to white, her shapely body turned to flab and, worst of all to Pee Wee, she had become "very religious." One aspect of the trip, however, provoked anger from Pee Wee. As he wrote to Atterton about the gig

> It was a little rough on Vic and Buck. It's a lousy segregation town. They were able to stay in the hotel but I eat with Vic in New York and nobody is going to stop me from eating with him in St. Louis. Nobody did. When Buck was nervous about cashing a check I really got sore. I went with him to get it cashed. And it got cashed. I don't know why I got on this, but it bothers me.

Next, Pee Wee played for three days with a local band in Fall River, Massachusetts, and then two days with another local band in Providence, Rhode Island. In May, he went to Chicago to record a technologically experimental album with Art Hodes' and Jimmy McPartland's bands. It was still the early days of stereo recording, and producer Jack Tracy had the idea of having one band play on the right channel and the other on the left. Pee Wee was in the Hodes contingent with Georg Brunis. Pee Wee hadn't seen Brunis in years and was surprised to see the trombonist's hair had turned "coal white," as Pee Wee described it. Pee Wee's aversion to Chicago had so deepened since his 1950 job with Hodes that he flew in for the recording date, then flew out on the first plane leaving afterward.

When he got home, he took a few weeks off from playing. With the acclaim he had finally achieved during the decade, he was able to pick and choose the dates he wanted to play. His health had improved greatly—he described himself to his old California friend, tenor player Slim Evans, as a "fat bastard"— but Mary began to complain regularly of stomach problems. She had a history of ulcers. Her doctor recommended an operation, but she decided against it, opting for a special diet and the elimination of cigarettes. Soon, however, she was smoking as much as ever.

Pee Wee had been looking forward to a projected recording for Riverside with Gerry Mulligan, but at the last minute the project fell through. Another that never materialized was an album featuring him playing with a string section.

> I've had an idea for several years that I'd like to record with strings, [Pee Wee wrote to Atterton] "just my horn and a bunch of strings. I tried to interest George Wein and he half promised to do it but nothing came of it. Recently Ruby Braff decided that I'm playing better than I've played in years and he got the idea (he thought) that I ought to record with strings. He talked to George Avakian about it. Anyway, it's something I've really wanted and I'm keeping my fingers crossed.

At one point, the prospect of doing a string album with some modern jazz musicians had been set, but at the last minute it was cancelled. "Mary says I'm more modern than any of them,"

Pee Wee wrote to Atterton, "because she never understood what the hell it was I am doing. I wanted that date. I made up my mind I'd scare those kids to death. Anyway, the hell with it."

At the 1959 Newport Jazz Festival, Pee Wee appeared with Buck Clayton, Vic Dickenson and Bud Freeman. Billed as the "Newport Jazz Festival All Stars," they were the first night's opening band. Ruby Braff, Count Basie's guitarist Freddie Green, and the great blues shouter Jimmy Rushing were added for a few numbers. After the concert, the front line went to Boston, where they recorded thirteen tunes that have been reissued on a multitude of labels under many different band names.

While Pee Wee was frequently hired as a featured soloist, steady night-club employement was scarce. The festival circuit then was nothing like it would become in ten or twenty years. Rock and roll had become America's popular music and, when the sixties began, jazz would submerge once again into an ocean of indifference. Pee Wee frequently would return from a job as a featured soloist to go for weeks without working at all. During that time, he'd settle in at the apartment and watch television.

Pee Wee returned to Chicago for the Playboy Jazz Festival on August 9, 1959, to play with Jimmy McPartland, Georg Brunis, Bud Freeman, Art Hodes and George Wettling in an "Austin High School" reunion band. In November, he played Stan Rubin's Thanksgiving bash again, in the grand ballroom of the Hotel Astor. December was spent again at Storyville in Boston with the "Newport Jazz Festival All Stars," An album resulted from the gig which included a slow blues, "Pee Wee Russell's Unique Sound."

The dawning of the sixties found Pee Wee in Philadelphia for a gig with Ruby Braff at the suburban Tally-Ho lounge. He returned there in March to play in Max Kaminsky's band, which included Bud Freeman, at Jack Field's Petti Arms lounge. Pee Wee's next significant date was on March 29, 1960, when he cut an album under his own name for Prestige Records with Buck Clayton. Much to the consternation of Buck and Pee Wee, the session turned out to be a dry one. They had planned to buy vodka on the way to the studio. Thinking Rudy Van Gelder's new recording facilities would be in a town, they drove to Englewood Cliffs, New Jersey, and finally located the studio in a wooded area, with no liquor stores in the vicinity. "No matter who we tried to get to go back to the city to buy some vodka," Clayton said, "we couldn't find one person who would go. Soon it was time to record, but we still didn't have a thing to drink. We even tried to get some beer but couldn't even get that, so we started to make the first song which was to be on the album." Clayton reported:

> We suffered. Pee Wee was so disgusted because it was his date and he had planned on having something there to drink. He grumbled about it as only Pee Wee could grumble, and I agreed with him because I guess this was the only time in our lives that we couldn't get anything to drink before a recording session. So we just knew that the record was going to be a flop because we couldn't get to our favorite booze. Then, just as we were preparing to make the last number of the set, someone came in with two bottles of hot beer and gave it to us. When the session was over, we both went home thinking that that was one session that was going to be actually nothing. A few weeks later, when we heard the finished record, we were surprised to find out that it was one of the best recordings that either of us had ever made . . . We couldn't believe that we had been cold sober when we made that recording, which shows you, I guess, that you can do something when you have to do it."

John Martin in *English Jazz News* wrote:

> The first impression that this record makes is that Pee Wee Russell is a poor clarinetist. The second reaction is—who cares? This is one of the finest records of uncluttered, free-wheeling jazz to come my way in the last two years. Russell's serpentine solos writhe stringily all over several well selected tracks. Now piping hot and unnervingly squeaky and shrill, now dark and

subterranean, his unique playing is the mark of an intuitive
musician rather than a master craftsman . . . I can't recall
ever having heard a trite chorus from Pee Wee in all my years of
listening experience and I am certain that it is this incredible
creativity that appeals to the many technically superior musicians who are always eager to work with him.

Stanley Dance wrote in *Metronome;*

Taste is important to both Pee Wee and Buck Clayton, and
basically this is therefore a good marriage. Pee Wee, with his
sincere approach, tortured lyricism, and ear for harmonies that
please and satisfy, seems to fall on the contemporary scene like
manna on the desert. That a cult now goes avidly for his *heart*
treatment is perhaps specifically due to the dearth of comparable
frankness and open handedness today . . . He has long been
appreciated in many quarters, but it required the current critical
climate for his talents to be reverently labelled as Art with a
capital.

Dance concluded that the record revealed a "Pee Wee uncaged."

It was as much the critics who were uncaged in their "discovery" of Pee Wee. Ever since his miraculous recovery in 1951,
he had been shedding his clown image. Now, thanks to Wein's
promotion, the jazz world was ready to take him seriously. The
suspicion of his being a charlatan had faded. The strange note
choices once thought wrong now were right. The beautifully
bent blue notes then condemned as off pitch now were appreciated for what they were. No longer put off by croaking vocal-
izations—they were used extremely infrequently now—jazz
fans began listening to the intricately constructed musical tales
of melancholy and yearning, of triumph and exhilaration.

Wein and a new bunch of "All Stars" opened at the Embers
in New York in April, 1960, the first time that intimate club had
a six-piece band. The band included Shorty Baker, trumpet;
Lawrence Brown, trombone and Bill Crow, bass. "That was a
wonderful band," Wein said. "You had the Ellington harmonies
with Lawrence and Shorty, and Pee Wee just fit in as only he

could." Wein wrote that "Pee Wee, as everyone knows my taste, is my favorite clarinetist. I don't believe he ever worked with Brown or Baker before, but the way these three veterans performed together was an experience in professional musicianship."

Crow said:

> I was used to hearing him [Pee Wee] in a more traditional setting, where all the horns in the front line improvised together contrapuntally. On this band, Lawrence fitted beautiful parallel harmony lines to Shorty's lead, leaving Pee Wee free to play whatever counter figures he chose without having to dodge anything else. His inventions were a wonderful surprise to us all, quite different from his usual ensemble playing.

Wein planned to bring the band to Boston following the engagement at the Embers, but Lawrence Brown was booked to play at another club there, so Wein recruited Bud Freeman to fill in. The band opened at Storyville on April 18, 1960, and the one week engagement was extended to May 2. On April 20 the band appeared on the local television program, "Dateline Boston: The Jazz Scene." They also played a couple of college dates.

On June 16, Wein featured Pee Wee and Baker with Tyree Glenn at a concert in the sculpture garden of the Museum of Modern Art in New York. Portions of the concert were issued on record a few months later to more critical acclaim.

> They treat . . . "I Ain't Got Nobody" to a soft-shoe tempo, rhythmic and comfortable, with a few old organ chords, and a superb solo by Pee Wee Russell who is, as you probably know, one of the great clarinetists in jazz [wrote Charles Edward Smith in *Metronome*]. "It seems too long between records by Pee Wee Russell, still blowing beautifully in all registers after forty years in jazz. This set brings him back in an informal session which also brings into focus the talent of Shorty Baker.

Pee Wee played the 1960 Newport Festival with Ruby Braff. In August, he appeared at the Quaker City Jazz Festival in Philadelphia with the Gene Krupa Sextet. In October, he was in Eddie Condon's band with Johnny Windhurst, trumpet and Roswell Rudd, trombone, at the London House in Chicago. Rudd, fresh out of Eli's Chosen Six, Yale's dixieland band, would become one of the leading avant-garde trombonists within a short time of this dixieland gig.

Pee Wee wrote to west coast tenor sax player Slim Evans:

> When I was at the London House with Condon, I made up my mind not to touch a drink, not even a beer. And I did it! Kept it up when we went to Detroit and didn't touch anything for weeks after I came home. And all I have is a little beer. Very little. But I have a helluva good appetite.
>
> I don't get around at all [Pee Wee continued]. I like nothing better than having my brew in front of the TV after a good dinner. I'm a changed guy. Whenever I think it might be old age I want to go out and have a few but then I decide it's not worth the effort. Anyway, I look and feel better as an old guy than I did when I was a young guy. And I think better.

In February, 1961 Pee Wee was reunited with Coleman Hawkins in a recording studio for another exceptional album. Nat Hentoff preserved some of the studio dialogue in his album notes:

> Nat Pierce, who had arranged all the numbers, wandered over and said to Hawkins: "Did you notice how that tune of Pee Wee's, '28th and 8th,' sounded like something Monk might have written?"
>
> "I can understand how it could," said the patriarch.
>
> "And some of Pee Wee's choruses," added Bob Brookmeyer, "are really way out."
>
> "I know, I know," said Hawkins. "For thirty years, I've been listening to him play those funny notes. He used to think they were wrong, but they weren't. He's always been way out, but they didn't have a name for it then."

To describe Pee Wee Russell's style with fresh adjectives, [wrote Charles Edward Smith in the *Metronome* review of the album] one would have to reach out into left field, where he seems to get some of his ideas. Often seemingly dissonant and disjointed in its rasping rhapsodic irregularity, it is not at all capricious, though these sorties have an air of being more spontaneous than his low register lyricism . . .

Nat Pierce remembered that they turned out the lights in the studio while he and Pee Wee recorded a slow blues, "Mariooch," one of Pee Wee's pet names for Mary, "to give us some atmosphere."

It may have seemed there were no new worlds to conquer, but after many years of thwarted attempts to tour Europe, Pee Wee finally made it in the spring of 1961. George Wein organized an April tour of the Newport All Stars built around the Essen Jazz Festival in West Germany. After that, he lined up a few more concert appearances in Berlin, Copenhagen and Paris.

The concert in Paris's famed Olympia Theatre was recorded and released a few months later. It had been held on April 15, the day the French army generals in Algeria revolted against Charles de Gaulle. "We gave two concerts that fateful day," Wein recalled, "and fortunately, the advance ticket sales for both had been good. It would have been a dismal evening for both the musicians and the promoters if this had not been the case. After noon on that day of revolution, not a single ticket was sold at the box office."

The European fans were passionate about Pee Wee's playing. "The French audience had never heard a musician like Pee Wee," Wein said. "The subtleties of his individualistic style, which have been ignored for so long by jazz fans in this country, were completely understood and enthusiastically accepted by the European jazz fans."

As excellent as Pee Wee's playing was during the tour, he was having a recurrence of physical problems. "At the time of the Paris gig," Wein recalled, "I went to his hotel room and there was blood all over the walls. He had been hemorrhaging.

We called a doctor. Pee Wee was amazing: he'd just recover. But he never really got over that illness he had in 1951."

Mary was also experiencing frequent discomfort from stomach pains. Despite their physical afflictions, they made the most of the trip, taking in the sights as much as possible, visiting many of the art museums.

In Berlin, the Newport All Stars played before 3,000 fans in the Sportspalast. The Feetwarmers, a German revival group, and British pianist Beryl Bryden were the opening acts. Bryden, who reviewed the concert for the April 26, 1961, issue of the English *Jazz News,* seemed more impressed by Pee Wee's trousers than by his playing.

> One of the first things that struck me about Pee Wee [she wrote,] was the razor-edged crease in his trousers. He looked just like a family lawyer or doctor. When he played he stood back from the mike, but the acoustics at the Sportspalast are so good that you could hear his every note and breath! He didn't play so many squeaks and grunts as I had expected, and his playing was sour and behind the beat. There was a nostalgic flavor about his playing that reminded me of Billie Holiday.

Despite Bryden's reservations about his performance, the trip was an unqualified success. Pee Wee and Mary only regretted not getting to England—that was postponed for another time. They had tested the waters and found that Pee Wee's renown was worldwide, his appeal universal.

12 • • • •

__"Modernist"

_R_eturning home to King Street, Pee Wee must have felt a great deal of satisfaction. The years of toiling night after night in Nick's or Condon's were far behind him. Now he was an internationally renowned jazz soloist, featured with only the best bands. He could pick and choose which dates to play. Almost immediately upon his return, he recorded with an all-star swing band for Prestige Records, billed as the "First Annual Prestige Swing Festival." The recordings were under-rehearsed and disappointing.

Pee Wee played at Condon's for several months in the spring of 1961 with Ruby Braff and valve trombonist Marshall Brown (who doubled on bass during the engagement). Condon paid proper respect to the clarinetist, no longer making him the butt of his jokes. Dan Morgenstern reviewed the group for _Metronome:_

> Braff and Russell are the outstanding soloists. Pee Wee's sound is, of course, one of the established identities in jazz. Today there is a mellowness which is not exactly new but more fully expressed. His truly beautiful two choruses on "Moonglow"— soft, full of gentle melancholia and hauntingly evocative of a cool September night—were the work of a true master.

Critic Stanley Dance, writing in *Metronome*, asked numerous jazzmen to fill out a wide-ranging questionnaire about their preferences. Pee Wee's responses demonstrated his sense of humor as well as his opinions. Asked to name his favorite "brand," he replied, "Miniature Schnauzer," a reference to his dog, Winky. He named Stravinsky his favorite classical composer, Rodgers and Hart his favorite songwriters, and Bix, of course, his greatest influence. The cornet was the other instrument he would most like to play, he said. He named two black vaudeville singers, Mamie Smith and May Alix, as the greatest blues singers he had ever heard. His favorite record of his own playing was the 1944 Commodore recording by his Hot Four of "Jig Walk." Finally, Dance asked, "If music were not your profession, which would you choose?" and Pee Wee responded, "I'd rather be dead."

In October, 1961, Pee Wee was trotted out to play the role of a Chicagoan again in the television special, "Chicago and All That Jazz," hosted by jazz fan and television personality Garry Moore. Pee Wee played with Jimmy McPartland, Jack Teagarden, Bud Freeman, Joe Sullivan, Eddie Condon, Bob Haggart and Gene Krupa. Although it was supposedly a gathering of old friends, the rehearsals were anything but convivial. All the old rivalries, rooted in long-forgotten incidents from several decades before, seemed to flare up. Condon, who was supposed to be in charge of the rehearsal, was hobnobbing, as usual, with everyone in the studio except the musicians, getting drunker as the day wore on. Teagarden, who regarded himself—rightly—as a star in his own right who had agreed to be a sideman for old times' sake, was especially incensed at Condon's attitude. The musicians managed to rehearse the numbers themselves. At the last minute, Condon came into the studio and, according to one musician, "started barking orders" to everyone. Teagarden, never known as a prima donna, began to walk out, but relented at the last moment after the entreaties of the others.

The musicians rehearsed, played on the Today television

show to give the audience a taste of what was to come on the special, and made a recording for Verve Records, all over the course of three days. It was touch-and-go throughout. Pee Wee was not too happy with the situation either. He resented more than ever being thought of as a Chicagoan, since he had been struggling to build a new audience for his playing. Here he was again playing "Royal Garden Blues" and "Indiana."

> They sent me a copy of the record the other day [Pee Wee told Whitney Balliett later]. I listened halfway through and turned it off and gave it to the super. Mary was here and she said, "Pee Wee, you sound like you did when I first knew you in 1942." I'd gone back twenty years in three hours. There's no room left in that music. It tells *you* how to solo. You're as good as the company you keep. You go with fast musicians, house-broken musicians, and you improve.

But John S. Wilson, reviewing the record for *Down Beat,* gave it four stars, noting that "Russell is simply magnificent in his tightly etched eloquence." *Newsweek,* however, in reviewing the television program, reported that

> neither hour nor circumstance suited Pee Wee Russell . . . who showed up on the color stage looking sad and baggy-eyed and feeling as blue as a flatted fifth. He moaned he hadn't slept in 48 hours. "Man, if you knew how I feel," he said, shakily trying to fit a reed into his horn.

Early in 1962, a year of momentous change for Pee Wee, he played at the Metropole Cafe with Jimmy McPartland's band. In March, he went on the road with George Wein's Newport All Stars, playing two weeks at the Theatrical Café in Cleveland and two weeks with Jimmy McPartland's band in Pittsburgh. "I shouldn't have gone to Pittsburgh," Pee Wee told Balliett. "I celebrated my birthday there and I'm still paying for it, physically and mentally. And the music. I can't go near 'Muskrat Ramble' any more without freezing up."

Marshall Brown began to play an important role in Pee

Wee's career around this time. The studious trombonist had been the band director at Farmingdale High School on Long Island. He and his youth band had appeared at the 1957 Newport Jazz Festival. The next year, he organized an international band of young professionals for the festival and the following year, the "Newport Youth Band." The critics were not kind to Brown or to his attempts to foster jazz musicianship among the young. Typical was John Hammond's observation about the International Youth Band:

> The only problem was that Marshall Brown was not the most inspiring leader in the world . . . They didn't like to be treated like high school students by Marshall Brown and there was fantastic resentment . . . All of them could out-blow Marshall . . . Marshall's heart was in the right place, but he was no inspiration to any real improvising jazzman.

It may seem strange that someone as individualistic as Pee Wee would team up with such a musician. Kenny Davern suggested that the clarinetist had formed a "Faustian pact" with Brown:

> Pee Wee despised being typed as a traditional player. Brown was an academician with money who wanted to get into the jazz world. He used Pee Wee to enhance his reputation in the jazz world. And Pee Wee on the other hand used the opportunity with Brown to do something different and challenging. It was an opportunity that rarely came his way because he was so typecast as a dixielander. It was a good premise for a relationship.

Pee Wee's interest in the association was more than that. It certainly is true that he was tired of playing the dixieland repertoire. He had scored a big success with his duets with Jimmy Giuffre and Tony Scott several years before and saw that, if he could escape being limited to employment in dixieland outfits, more jobs would be his. But beyond that, although Pee Wee did not have a strong self-image, he had a strong sense of who he was musically. He felt his style, so long regarded as weird, laughable, even suspect in a dixieland setting, would sound very contemporary in a modern band. He realized long before others that the context Brown could provide would be perfect for his unique harmonic excursions.

The two started rehearsing with Russell George, bass, and Ron Lundberg, drums, in Brown's home studio. Their repertoire was decidedly modern, even avant garde, featuring compositions by Tadd Dameron, Thelonious Monk, John Coltrane and Ornette Coleman. Brown taped many of the sessions to assist him in "helping" Pee Wee. Brown saw the relationship as a teacher-student, and at first Pee Wee managed to take the instructions in good spirits. "We instigated that," recalled Davern, "Roswell Rudd and I. We used to go to the rehearsals in Marshall's studio. The fact is that we said, 'Yeah, Pee Wee, that's what you should be doing. That's the context which you can be heard the best in.'"

In March, 1962, Pee Wee played a unique gig: as best man at Kenny Davern's wedding to Sylvia. When the *Melody Maker* ran a photo of Pee Wee at the ceremony, Mary cut it out and mailed it to the newlyweds, noting that it was the first time the best man had upstaged the groom.

In April, *New Yorker* critic Whitney Balliett came by the Russells' apartment and interviewed Pee Wee for a "profile." The piece would be published in August with the title, "Even His Feet Look Sad," and would give another boost to Pee Wee's career. It was the finest biographical piece on Pee Wee to appear during his lifetime. Pee Wee talked to Balliett about the new quartet he and Brown were rehearsing:

> When I was sick, I lived night by night. I was bang! straight ahead with the whiskey. As a result, my playing was a series of desperations. Now I have a freedom. For the past five or so months, Marshall Brown and I have been rehearsing a quartet in his studio. We get together a couple of days a week and we *work*. I didn't realize what we had until I listened to the tapes we've made. We sound like seven or eight men. Something's always going. There's a lot of bottom in the group. And we can do

anything we want—soft, crescendo, decrescendo, textures, voicings. What musical knowledge we have, we use it.

A little while ago, an a & r man from one of the New York jazz labels approached me and suggested a record date—on his terms. Instead, I took him to Brown's studio to hear the tapes. He was cool at first, but by the third number he looked different. I scared him with a stiff price, so we'll see what happens. A record with the quartet would feel just right. And no "Muskrat Ramble" and no "Royal Garden Blues."

Changes also were happening in Pee Wee's domestic life. He and Mary were finally able to move from their King Street apartment to one they had seen (more than a year before) at 345 West 28th Street. The new building was owned by the International Ladies' Garment Workers Union, and Pee Wee and Mary had been on "the list" for more than a year. Although an international success, by choice Pee Wee had not been working much that year, and Mary was supplementing their income by working in the statistics and advertising department of Robert Hall Clothes. The extra money meant new furnishings for the new apartment.

Another source of funds for the Russells came from Mary's nephews, Sol and Lee Goodman, who were very supportive and sent money regularly. The Goodmans, who had prospered in the diamond business, liked and respected Pee Wee. The close-knit family was proud of its "bohemian" aunt and "artistic" uncle. Furthermore, they genuinely appreciated Pee Wee's talent and the importance of his role in the history of jazz. The Russells were frequent guests in their homes, and their children eagerly awaited visits from "Uncle Pee Wee." When Pee Wee would visit Lee and Estelle Goodman in Union, New Jersey, the men would often sneak off for a bottle of ale at the local saloon. "Mary would always know he had had an ale from his eyes," Goodman said. "She would give him hell." The new apartment was more or less a gift from the Goodmans.

"I took over the old place on King Street—the apartment cost

$55.25 a month," said Kenny Davern. "They told the landlord that if I couldn't have it, they wouldn't move." Davern had become very close to Pee Wee by then:

> They left everything behind including their dog, Winky, except for the television set. It was one of those ancient eight by ten inch sets in a huge box. Roswell Rudd and I got a station wagon and carried it over to them. A few days later, Pee Wee came over to see if there was anything else they wanted. While he was looking, I accidentally tipped over his recliner. Under it, we saw something that looked like it had been alive at one time. It is a formless shape, covered in grime and dust—Mary never cleaned. We poked it with a stick and finally I picked it up and wiped off years of accumulated dirt. It was a clarinet case and in it was a Conn. Pee Wee looked at it and said, "So that's where that went!" He had lost it eight years before. He insisted I keep it and I still have it to this day. [The clarinet was a rubber Conn, model 324, serial number 268143.]
>
> They couldn't have a dog in their new place. Sylvia and I had Winky for two weeks when Pee Wee and Mary came over to visit their "daughter." The buzzer rings, and Winky starts to bark. Mary and Pee Wee walk in, and Winky's sitting on a chair. Pee Wee goes over to greet Winky, and the dog runs away. "Why, you turncoat," Pee Wee says, as the dog scampers into the next room. Winky would have nothing to do with him after that.

Pee Wee was one of the Newport All Stars at the festival on July 8, 1962, with Ruby Braff. Whitney Balliett noted in *The New Yorker* that Pee Wee's four low-register choruses on the blues were "just as good as any of the countless uncanny blues solos he has tossed to the winds for decades." Louise Tobin, who had been a vocalist with the Benny Goodman orchestra in 1939, sang a few numbers with the band, and Bud Freeman was added for a few tunes. The band's last number was "Take the 'A' Train," on which they were joined by the Duke Ellington orchestra. The event was filmed and released as part of the feature-length *Newport Jazz Festival—1962.*

Later that month, on July 25, Pee Wee appeared with Ruby

Braff and Dickie Wells at a benefit concert for reedman Eugene Sedric at Central Plaza. The band also accompanied vocalist Esther Williams, daughter of the 1920s band leader, Fess Williams, on several numbers.

On August 26, shortly after Balliett's profile appeared, Pee Wee appeared with the Newport All Stars at the Ohio Valley Jazz festival in Cincinnati with Ruby Braff, Marshall Brown and George Wein. A city commissioner, confusing jazz with rock and roll, attempted to halt the event, fearing the grandstand at Carthage Fairgrounds would not withstand the "excessive foot stamping" he associated with jazz audiences. The issue became, according to Wein, a *cause célèbre*, and "the fight between rock and jazz took over."

More than 17,000 people attended the three-day event—and resisted stamping their feet. The All Stars played the second night, August 25. They were barely into their first number, "Keepin' Out of Mischief Now," when torrential rains came. Audience and musicians scampered for shelter. The following night, the band returned. "Russell was excellent on a slow blues and a fiery 'Indiana,'" wrote Don DeMichael in *Down Beat*. Later in the evening, Russell and Braff sat in with the Jack Teagarden Sextet, but what could have been a stimulating set was cut short. After the festival, the band played at the Surf Club in Cincinnati.

Meanwhile, Brown was forging ahead with plans for the debut of the quartet. He and Pee Wee selected several performances from the rehearsal tapes and pressed one hundred copies of a "demonstration" disc. Brown sent it to prospective nightclub owners and eventually landed a gig at the Town Tavern in Toronto. A few days before the gig, however, George Wein recorded the Newport All Stars for Impulse. The album included another blues composition by Pee Wee, named by Braff "The Bends Blues" because of the way that Pee Wee bent notes—and his body—during the performance.

Finally, on October 15, 1962, more than nine months after

they started rehearsing, the Pee Wee Russell Quartet with Marshall Brown, as it was billed (with George and Lundberg), made its debut in Toronto. According to Helen McNamara in *Down Beat*:

> While it may take a while for the wistful cries of Russell's clarinet playing, contrasted againt Brown's powerful and facile playing of valve trombone and bass trumpet, on ensemble passages they were most effective, whether they were aiming for a velvety whisper of a sound on slow romantic ballads or a big, fatty blare on the up tempos. The sheer novelty of the new Russell group will be a big attraction. It may send some of the early-day Russell fans out into the night muttering about the nerve of Charles Ellsworth forsaking dixieland, but it could also point the way to a greater use of the talents of the jazz giant in contemporary settings . . . Whatever the outcome, it was worth every moment to see the pleasure with which the clarinetist attacked every tune during the quartet's debut here. It was almost as though he were making up for each time he has had to play "Muskrat Ramble."

Not all the reaction was as positive. Reviewing Pee Wee's appearance at the Town, Patrick Scott wrote in the *Toronto Globe and Mail*:

> Mr. Russell is doing exactly what he has done best for 40 years now, which is to play the clarinet like Pee Wee Russell. The fact that he is doing it at the Town, instead of at Nick's, and with an accompaniment and a repertoire that would drive his former colleagues to drink, is so much superfluous detail—although it may also be part of the most cunning hoax in jazz since the doors first opened at the House of Humbug.
>
> If Mr. Russell's clarinet has an even odder sound than usual these nights, I suspect it is the result of his playing with his tongue in cheek; either that or he really does believe what he has been reading about himself, which, since he has an acute sense of humor, I find hard to believe. His recent acceptance by the modernists—whose predecessors during the 1940's made him the primary target of their ridicule and scorn—does not alter the fact that his playing this week on Mr. Coltrane's Red Planet

differed by not so much as a single squeak from his playing 35 years ago on "Feelin' No Pain."

Indeed, many of Pee Wee's fans were upset, not only at the sound of the quartet, but about his pronouncements about the music with which he so long had been associated. The flirtations with modern jazz, his duets with Giuffre, had been excused by purists as another one of his harmless eccentricities, but now he was heard to be damning the very music that had championed him for so many years. To *Down Beat*'s Bill Coss, he complained;

> I felt I wasn't doing anything. I could read a newspaper while I was playing a chorus. You get stagnant. You stand still. You get old before your time. I have nothing against Dixieland—not that I ever played it—but there's no challenge. How many records can you make of "Sugar"?

In an interview in the *Boston Traveller,* Pee Wee said;

> I'm still searching. I like to experiment, to try other things . . . I never played Dixieland in my life. I played jazz music, that's what I played. Any kind of music that's played well is good . . .

Asked by jazz writer Les Tomkins if he had set out to prove his musical scope was wider than some might have imagined, Pee Wee replied:

> I've been trying to prove it to *myself* once in a while—that I'm still alive. That's all. So I do a little quartet thing. So we play some Coltrane and some Monk. I just wanted to see if I could do it. To satisfy myself that I wasn't going to sleep. I'm satisfied, so that's that. So something else comes along and I'll try that, maybe. I don't say it's good. I don't ask anybody to like the crazy noises I make. But I've been living on 'em. I'm 58 and I'm still getting along. I'm sleeping inside and I always will.

Through the efforts of Jeff Atterton, Pee Wee's English pen pal who had relocated to New York, the group was offered a contract to record for Columbia Records. Atterton saw to it that

Pee Wee received the biggest advance of his career for making the recording—$1,000. The sessions began on November 12 with a recording of "Pee Wee's Blues." They continued through December 4, 1962. *Jazz* magazine described the scene in the studio:

> Ron Lundberg arrived early; somewhat tense for his first record date . . . Pee Wee arrived and busied himself away from the others, carefully putting his instrument together. He opened the bag he had brought with him and extracted several bottles of ale . . . he looked very healthy. He's been taking care of himself.
>
> The group [had] rehearsed for several months. They ran down "My Mother's Eyes." One is aware that this is not a run-of-the-mill blowing session. Brown set the pace. They do two excellent and acceptable takes. Brown is not worried. They do a third take, "just for safety."
>
> The playbacks are a mixture of reactions. Brown is ecstatic. Pee Wee looks concerned. More presence is needed on the drums. Brown says, "We'll do one more!" They complete another take. Brown tells the others to knock off. The session is running smoothly, according to plan.

But all was not well with the quartet or the recording sessions. Brown insisted on editing the tapes to eliminate his mistakes. It got to the point that the producer, Frank Driggs, was ready to throw up his hands and walk out. And Pee Wee was smarting under Brown's martinet manner:

> It was challenging for Pee Wee [said Atterton], tough for him. Marshall was a taskmaster. He leaned on Pee Wee to extend himself a little. Pee Wee wasn't used to playing charts. Pee Wee did get mad once or twice during the sessions. He was impatient with the discipline Brown insisted on. There was a little verbal argument. Marshall wanted to try and get the thing a little tighter and Pee Wee didn't want to.

The album, issued as "New Groove" by Columbia, was met with critical raves.

Russell comes of age [wrote *Esquire*], completely abandoning the dixieland formula which has restricted his tremendous restless talent. In this imaginative new album, his pulsating clarinet, paired with Marshall Brown's valve trombone, Russell George's bass and Ron Lundberg's drums, produces a bracing set of fresh jazz sounds. The range and variety of the album is demonstrated by his choice of tunes . . . He even turns the maudlin George Jessel's "My Mother's Eyes" into an arresting jazz piece. The overall ensemble effects create sounds that have the feel of a much larger group. A consistently stimulating, creatively superior jazz disc.

Despite the great reviews, sales failed to materialize. Less than 10,000 copies of the album were sold. Many of Pee Wee's old-time fans were dismayed by the modern experiments, while young turks still tended to regard him as a relic of the past. Nevertheless, the critical acclaim resulted in Pee Wee's winning the clarinet category in both the *Down Beat* international critics' poll and the *Melody Maker*'s critics' poll, the first such recognition he had received since the early 1940s when he appeared regularly on the popular Condon Town Hall concert broadcasts.

In *Down Beat*'s yearbook for 1962, critic John S. Wilson displayed remarkable insight in an article entitled "The Strong and Vigorous Mainstream":

One of the most remarkable fish caught in the mainstream net is Pee Wee Russell. During the days when most of the others now classified as mainstreamers were flourishing in big bands, Russell worked the relatively limited small-group field. And, things being what they were then, that meant that he was usually in dixieland surroundings.

But, as Russell had frequently protested in recent years, he is not now and never has been a dixielander. Certainly the strange collection of squawks, squeaks and marbled moans from which he builds his solos have little relationship to dixieland. Yet, lacking any other means of identification, he was lumped with the dixielanders until, in the last few years, it had begun to be

evident that the problem lay with the times, that jazz was just
catching up to him in the 50's, and that Pee Wee has done a
remarkable thing: in the 30's and 40's he managed to be both
avant-garde and popular.

Although record sales were disappointing, the quartet found
enthusiastic audiences at the Latin Casino in Cherry Hill, New
Jersey, where they played for a week at the end of December,
1962. In April, 1963, they cut a second album, under George
Avakian's supervision. The tapes were sold to Impulse! and issued under the title "Ask Me Now." The total cost of the recording was $2,401.65, with $848.05 allocated for payment to the
four musicians.

Sinclair Traill, reviewing "Ask Me Now," in the September,
1966, issue of the English *Jazz Journal,* reported,

> When Pee Wee said to me last year that . . . he didn't like the
> playing of Marshall Brown . . . and that the whole thing—he
> was speaking about the first album issued in 1963—was too
> carefully scored, he was saying just what I think of this record.
> He went on to say that "to get any kind of group sound and
> balance, one cannot rely on the written note—it has to come
> from listening to what is being played around you. If you don't
> listen that way, then you are left 'naked'—you hit a note but you
> are all alone." And that again exactly fits my feeling towards this
> LP.

It was only fitting that Pee Wee's modernism be on display at
the Newport Jazz Festival in July, 1963, when he was featured
with Thelonious Monk's group. Mary and Pee Wee arrived at
the festival a week early as house guests of George Wein, who
had suggested the idea to both musicians:

> I told Monk that Pee Wee had the same feeling for intervals that
> he had. They both understood that intervals are what give you a
> style that other people don't have. That's something Benny Good-
> man never paid any attention to. As great a clarinet player and
> musician as he was, he hadn't worked it into his style.

In the days before the concert, Pee Wee and Mary enjoyed the Newport countryside. "I was here at the beginning," Pee Wee reminisced, "when we didn't know if people would go for a festival or not. Money was no problem but we didn't know if jazz would draw enough people to make the effort worthwhile." Mary thought Newport was a wonderful town. While shopping for ale for Pee Wee on Bellevue Avenue, she found everyone "so courteous. Cars stopped for us when we crossed the streets. In New York they would run over you," she said.

Critical reaction to the pairing of Pee Wee and Monk was excellent, although Pee Wee was not impressed.

> The combination of Pee Wee Russell and Monk on Thursday [wrote Ira Gitler in *Down Beat*], turned out very well. Russell, who had picked the tunes he wanted to play at Newport while listening to Monk at New York's Five Spot but had not been able to rehearse them with the group, found beautiful passages in "Nutty," but he was over-shadowed by a great, intense Rouse [tenor saxophonist Charlie Rouse] solo. In "Blue Monk," however, the clarinetist communicated very well through his personal poetry, cast, this night, in Monkish mold.

New York Times critic John S. Wilson had some reservations:

> Mr. Monk did not play while Mr. Russell was playing and Mr. Russell did not play while Mr. Monk was playing so that the meeting amounted to little more than their presence on the same platform at the same time . . . this meeting failed because they never got close enough to make contact.

Not so, countered critic Dan Morgenstern in *The International Musician*.

> Contrary to published reports, Monk and Pee Wee did play together; in fact, Monk treated his guest just like a member of the tight family group. Pee Wee's most eloquent solo came on "Blue Monk," where he played his own special brand of blues, yet built his solo solidly on Monk's theme, like a real Monk player.

Both Wilson and Morgenstern were right. Monk did play behind Pee Wee's first solo chorus on each tune, then, as the pianist did with Charlie Rouse's subsequent choruses, he sat out leaving the reedman accompanied only by bass and drums. However Monk's accompaniment was limited almost entirely to comping behind Pee Wee: the pairing set off no sparks. When one of the tracks from the concert was played for Pee Wee during a blind-fold test in 1964, his reaction showed that he had second thoughts about the whole modern experiment. Of his perfor-mance with Monk:

> Well, I don't think that should ever have been issued. No re-hearsal, just pushed onto the stage, and I didn't fit into that group. Anyway, I don't like that kind of music. I know a lot of critics said that thing they called "New Groove" we made a couple of years back showed that I had become a real modernist. That wasn't right. There was a whole lot about that record I didn't like. To start with I don't like the playing of Marshall Brown on that record. He wrote all that stuff out and rehearsed us every carefully, but had it been played in any other style, then he couldn't have done it all. It was very hard to do, but excepting my solos, it was all scored.

The group—now expanded to a quintet with the addition of pianist Bob Hammer (with Jack Six on bass and Bedford on drums)—played their last engagement at the Village Vanguard for a week in early September, 1963. *Newsweek* reviewed the performance.

> Russell appeared in New York last week at the Village Vanguard at the head of a new quintet. Marshall Brown announced the numbers. "Pee Wee doesn't like to talk," he explained later. For about a year, he and his group had played as a quartet—without the added piano—in such cities as Toronto, Philadelphia and Camden.

Pee Wee's transformation into a modernist seemed an espe-cially astute career move in August when the venerable Nick's passed almost unnoticed from the scene. The *New York Times* reported that Mrs. Rongetti had lost the lease and was closing the doors. The whole culture that had been so vital from the middle 1930s through the end of the war had faded away.

"Chicago music has been dead for years," *Newsweek* quoted Pee Wee as saying. "It means nothing. It is a live recording of the deep past, but it is like clowning . . . All my life I've been a lone wolfer. They used to say, 'What the hell are you doing?' and I would say, 'This is the way I want to play.' Playing with younger, mainstream muscians," concluded *Newsweek,* "he fits in smoothly and sounds like a contemporary jazzman with a future, not like an aging echo of the past."

Pee Wee's association with Marshall Brown ended with the Vanguard job. Brown's didactic manner rubbed Pee Wee the wrong way. "Pee Wee wants to kill him," said Mary. Pee Wee said, "I haven't taken so many orders since military school." Another reason the partnership went sour was that, although Brown had indicated to Pee Wee that he could raise the money to keep the group together, he did not come up with it when it was needed. Pee Wee's inclination to perform in a modern setting was correct, but his choice of Brown was wrong. Brown not only did not swing; he had a poor sense of time, and the young rhythm section was not strong enough to compensate. Pee Wee should have sought someone whose arrangements and playing would have stimulated him. Someone like Gerry Mulligan would have been ideal.

Pee Wee went to Boston's Tic Toc Club, where he fronted a local band for a week. He was interviewed there about his career.

> I'm only starting to feel really well again now. People ask me what my hobby is. It's music. Watching television and playing golf [it is doubtful that Pee Wee ever played golf in his life] are pleasant pastimes, but if I had to do that the rest of my life—I'm 57—I'd just shrivel up and blow away. I've seen others—you know, give me the simple life—I've seen them do it. If I stopped music, I'd be dead.

Asked about the lingering divisions in the jazz world, Pee Wee said: "There's absolutely no cause to ride anybody. It's a sign of stupidity to blame and hate other people for what's happening to you," he said. "Now, being tough, that's something else. I did tent shows when I was kid. I went through the mill, and I mean it. I know the best sides of life and the lowest—I got through it when I was young. You have to use toughness at the right time and with discretion. You've got to know when and how . . . Formal musical foundation is good to have, but you've got to get kicked around too. Someday you're going to meet somebody who knows so much more than you, you'll be scared stiff."

During the engagement, Pee Wee called up Robin Goodman, daughter of Mary's nephew, Sol Goodman. Robin Goodman (now the producer of the Second Stage Theatre in New York) was attending Brandeis at that time, when the sixties were getting under way in earnest. She was especially close to "Uncle Pee Wee." "He was such a pure artist, the first artist I ever met," she said. Uncle Pee Wee asked her if she was going to come and hear the band. When she answered in the affirmative, he asked her if she could get an "ounce of grass" and bring it with her. "It's not so much for me," he explained. "It's for the boys in the band." Robin complied with his wishes.

Following the Boston engagement, Pee Wee is reported to have gone to Canada to make a pilot film for the Canadian Broadcasting Corporation, but nothing more is known about this. On September 12, 1963, he appeared at the Foxborough Race Track with a nine-piece band that included tenor sax star Zoot Sims. Then Pee Wee flew to Cape Cod, where he was featured in Bobby Hackett's band for a Sunday jam session. Eddie Condon also was hired for the job. Pee Wee had another offer to play that night for more than twice the money. He asked Condon what to do and the guitarist advised him to take the Hackett gig. "Think of your poor ears," Condon said. On the bandstand, Condon took over as master of ceremonies, regaling the audience with stories at Pee Wee's expense until finally the clarinetist grabbed the microphone from him and told him to either talk about their leader, Hackett, or shut up. Pee Wee had come a long way since the days at Nick's when he suffered Condon's jokes in silence.

Next, Pee Wee flew to California where he appeared at the Monterey Jazz Festival with Jack Teagarden and an all-star band, on September 20, 1963. It was the first time he had been back to the San Francisco area since his near-fatal illness in 1951. The group originally was to be billed as the Teagarden-Russell All-Stars, but by the time Pee Wee arrived the entire affair had turned into a Teagarden family reunion, with Charlie Teagarden on trumpet and Norma Teagarden replacing Joe Sullivan at the piano on several numbers. Even "Mama" Teagarden appeared and played a couple of rags on the piano. Bassist Jimmy Bond and drummer Nick Ceroli, who appeared with Gerald Wilson's big band just before the All-Stars took the stage, filled out the band. Gerry Mulligan also sat in on baritone sax for several numbers, including a long blues with Pee Wee. Don DeMichael, reviewing the set for *Down Beat,* said, "Pee Wee Russell was not at his best, though a blues duet by the clarinetist and Mulligan, despite a what-key-are-we-in? beginning, was charming." Mulligan recalled

> We were really doing an Alphónse and Gaston, treading lightly so as not to step on each other. What came out of it was such a gentle piece of music. It is rather remarkable.
>
> He was inclined to be further out—harmonically and melodically—than I am. I have always based everything I've done on conventional harmony. Pee Wee was unique. He was fearless. I never thought of him as a clarinet player—it was more like a direct line to his subconcious. Pee Wee was a presence.

During the festival, Jack Teagarden contracted the flu and was running a temperature. Nevertheless, he went on with the show, singing and playing his usual repertoire as only he could. While he was there, Pee Wee participated in a symposium, "What Ever Happened to Dixieland?" and appeared with the band again the following afternoon, without Mulligan and with George Tucker replacing Bond on bass. "The set was highlighted by more excellent Teagarden trombone and better playing by Russell and Charlie Teagarden," wrote DeMichael.

Immediately after the second concert, Teagarden went into a hospital. His regular band was scheduled to perform at Curley's Lounge in Springfield, Illinois, but when it became evident Teagarden was too sick to play the ten-day gig, Pee Wee agreed to fill in, with a local rhythm section. It was the last time Pee Wee saw his friend from the Peck Kelley days, nearly forty years before. Within five months Teagarden was dead.

Pee Wee capped 1963 by again winning the *Down Beat* critics' poll. Whatever the problems had been with Marshall Brown and with Thelonious Monk, Pee Wee had successfully made the transition in the critics' minds from "living legend" to "modernist."

Pee Wee started 1964 with preparations for a tour with Eddie Condon's All-Stars of Australia, New Zealand and Japan, scheduled for the spring. He asked Kenny Davern if they could practice some technical exercises together.

> Pee Wee was going to Japan and he thought he might have to "cut a show," and he wanted to brush up on his reading [Davern remembered]. He hadn't read in such a long time. We met at a little studio that was run at that time by Jake Coleman called Bob Jewel's studio, on West 48th Street. I pulled out a couple of very simple Mozart duets that had been transcribed for two clarinets. They weren't anything harder than dotted sixteenths. He played it just as he would have played a jazz chorus.

Before the tour started, Pee Wee played an engagement with the Jewels of Dixieland at Bovi's Town Tavern in East Providence, Rhode Island. At the end of February, rehearsals started for the tour, for which Condon had brought Pee Wee together with Buck Clayton, Vic Dickenson, Bud Freeman, Dick Cary, Jack Lesberg and Cliff Leeman. Kansas City blues shouter Jimmy

Rushing was the vocalist. The first stop for the band was San Francisco, on March 1, 1964, where it played two nights to full houses at Turk Murphy's "Earthquake McGoon's." During the gig, Pee Wee bunked at jazz writer Dick Hadlock's house in Berkeley.

The Australian tour was produced by Kym Bonython. The band arrived in Sydney on March 5, then flew on to Adelaide. The Australian jazz musician and leader Graeme Bell met the gang at the airport. "You are very well known in America," Condon said. "But tell me, just how did you get that idea to invent the telephone?" That night, they were interviewed on television. The following day, they were interviewed by Bonython on his radio program, Tempo of the Times, and that night played the first of four concerts at the Adelaide Festival of the Arts in the Regent Theatre. The band caused a sensation, and Condon's witticisms were quoted for days in the Australian press. Pee Wee was a special favorite of the fans. His featured numbers, "Sugar" and "Pee Wee's Blues," never failed to bring down the house.

Dick Hughes, an Australian jazz pianist, critic and raconteur, who acted as road manager for the musicians, wrote in his autobiography, *Daddy's Practicing Again,*

> Sugar was possibly the greatest number of the tour. It was some of the most beautiful music I've heard in the flesh and moved me as much as Walter Gieseking playing the Debussy Preludes in Melbourne Town Hall twelve years earlier. Pee Wee breathed into the clarinet and monumental music was created.
>
> Funny thing happened to Pee Wee on his way to the microphone for his first solo at the first concert. Pee Wee edged up to the microphone, gave his conspiratorial wink to Dick Cary—and then a loud voice came from the audience, "Don't get nervous."
>
> It broke up everybody, but not, for some time, Pee Wee, who looked out at the audience, rapt in wonderment. He tugged his ear, put his head to one side, looking like a perplexed budgerigar [an Australian parakeet or "budgie"]. Then he grinned, blew a

quiet blues chorus, then seemed to stagger back from the microphone a pace or two to open up his heart in some of the most
lyrical, anguished blues I've ever heard. Some of it was modern
enough to have come straight from a contemporary Thelonious
Monk composition; some seemed to arch back to the country
blues wailers like Blind Lemon Jefferson and Lightnin' Hopkins.
In three or four minutes, Pee Wee gave us the very essence of
jazz and blues and said more in that time than some of the way-
out boys hint at in forty-five minutes.

After the festival, the band moved on to Melbourne, where they
played concerts on March 11, 12 and 13 in Festival Hall. Then
the band went to Sydney for two concerts at Sydney Stadium.
Hughes' band opened the concerts in Sydney.

Part of the band played three shows at the Playboy Club in
Melbourne on March 15, but Pee Wee, Condon and Jimmy
Rushing had stayed behind in Sydney for the day. That night,
everyone flew to Auckland. The All Stars appeared in ten concerts in New Zealand, one each at six and 8:30 p.m., on Monday, March 16, at Auckland Town Hall, March 17 at Hamilton
Embassy Theatre, March 18 at Wellington Town Hall, March
19 at Dunedin Town Hall, and March 20 at Christchurch Civic
Theatre. The contract with New Zealand Broadcasting Corporation, which arranged for the tour, included rights to record
the concerts for broadcast. The *New Zealand Press* in Christchurch wrote:

> What an extraordinary talent Pee Wee Russell has. Over the
> years his playing has taken on a kind of classicism; it now has the
> timeless quality of the best in any art . . . it was a magnificent
> performance, beautifully balanced and developed.

On March 23, the band arrived in Japan. They appeared in
concert at the Hibiya Civic Auditorium in Tokyo on March 24
and 25. From March 26 through April 1 they toured Osaka,
Nagoya, Sapporo and Kyoto. They returned to Tokyo on April 2
and performed at Sankei Hall. The concerts were all sellouts

and the crowds were very enthusiastic. The first Tokyo concert was recorded by the Japanese Broadcasting System and portions were later issued in the United States and Japan.

The weary All Stars left Japan on April 3. Pee Wee's lyrical, imaginative playing had produced an outpouring of admiration and respect in all three countries. The tour was an unqualified triumph for him. Plans fizzled for a return engagement at Earthquake McGoon's on the way back from Japan. The Condon All-Stars headed home, where more jobs awaited Pee Wee. "That was the most back-breaking trip of my life," he wrote to Slim Evans. "I slept for ten days straight when I came home."

On June 20, 1964, a refreshed Pee Wee was featured at the Pittsburgh Jazz Festival, sponsored by the Pittsburgh Catholic Youth Organization, in the civic auditorium there. During July he was a member of a Newport All-Stars unit (with Max Kaminsky, Bud Freeman, Bob Hammer, piano; Bill Cronk, bass; and Cliff Leeman, drums) at the Metropole Cafe in New York. During the gig, the Metropole put on a mini-jazz festival of its own, with two groups of Newport All-Stars and a special appearance by Louis Armstrong and His All Stars. The event broke all attendance records at the club, with block-long lines of fans waiting to get in.

There was a "Tribute to Eddie Condon" held at Carnegie Hall shortly after midnight, July 21, with Sammy Davis, Jr., scheduled to be the master of ceremonies. But Davis, who was appearing in Philadelphia in *Golden Boy,* said he was too fatigued to make the trip to New York and cancelled at the last minute. The affair was staged to help pay for the three prostate operations Condon had undergone since his return from the Asian tour. Condon was not happy with the event, which featured Bob Crosby's band and Woody Herman's Thundering Herd. For a man who had spent his career in opposition to the big bands, the choices were not appropriate and put Condon, who had been drinking heavily, into a surly mood.

Some of the figures Condon had been associated with through the years were on hand, however, including Johnny Mercer, Bud Freeman, Wingy Manone, J. C. Higginbotham, Willie "the Lion" Smith, Jack Lesberg, Cutty Cutshall, Gene Schroeder, Bobby Hackett, Peanuts Hucko, George Wettling and Bily Butterfield. The high point of the concert was when Henry "Red" Allen sang and played "I Ain't Got Nobody" with a particularly effective backing by Pee Wee. "What Eddie proved," Pee Wee told *Newsweek* at the concert, "is that our music has vitality—it's alive. I've played both; this has more feeling." Clearly, following the Marshall Brown episode, Pee Wee was rethinking his antagonism to dixieland. The concert formed the basis for a television "tribute" which was recorded several months later with Davis as master of ceremonies, but without Pee Wee.

"I'm a busy Indian," Pee Wee wrote to Slim Evans.

> I leave tomorrow to join Hackett on the Cape for a week. From there I do a single in Toronto for two weeks. Home for a few days and then some one-nighters. Then the Monterey jazz festival. Home for two days and then off to Europe. Paris, Copenhagen, Helsinki and a lot of others. When I'm finished with that tour (a George Wein package) I'm being booked for a single in Great Britain. It's all set except for the exchange. If they can't get a guy into the States to exchange for me it's all off, but if they can get an exchange, I'm set for three weeks in England and maybe Scotland.

From August 10 through August 22, 1964, Pee Wee played in Toronto at the Colonial Tavern with Jim McHarg's Vintage Jazz Band. The Colonial's owner, Good Lichtenburgh, gave McHarg, who had recently emigrated from Scotland, his first "break" by hiring his band for two weeks. They backed trombonist Lou McGarity during the first week and Pee Wee during the second. McHarg said:

> We were thrilled to appear on the same stand as Pee Wee. He was easy to play with. He was a quiet man, almost timid. I think

he knew that we worshipped him and we were all scared to death that we didn't measure up. I know he really enjoyed his week with the band. From my point of view, it was a highlight of my career.

One incident showed a different side of Pee Wee. I picked him up at his hotel one day to go down to the gig. As the taxi took off Pee Wee went a bit hairy and screamed to the taxi driver that he was going the wrong way. It almost seemed as if it had something to do with personal survival. When the driver made an agitated U-turn, Pee Wee settled back with a self-satisfied grin. I couldn't believe it. He was completely out of character.

Pee Wee was notified that he had won the clarinet category in the *Down Beat* international critics' poll once again, with 88 votes. Jimmy Giuffre was second with 37 votes. Upon his return to New York, Pee Wee rejoined the Newport All Stars for a performance, then left for the Monterey Jazz Festival, fronting his own band (Buck Clayton, Vic Dickenson, Bud Freeman, Dick Cary, Red Callender, and Earl Palmer, drums), playing opening night, September 18, 1964. Gerry Mulligan sat in with the band, and Count Basie's singer, Joe Williams, was featured on a couple of numbers. "Steals Show on Opening Night," headlined a review of the band by Fred Hill:

> The night clearly belonged to a convivial group of jazz artifacts led by Pee Wee Russell. The Russell group played a rousing opener, then Cary was featured on alto horn on "Caravan," "Basin Street" (Dickenson feature), "'S Wonderful" (Freeman showcase), "Lullaby of the Leaves" (Mulligan), "Stompin' at the Savoy" (Clayton), then "Pee Wee's Blues."
> The last to display his talents was Pee Wee, playing "Pee Wee's Blues," which began with a plaintive blues strain, erupted in the middle into a violent "dirty blues," and subsided again into the opening phrasing. It was everything that jazz should be and the applause was fitting. Russell played one more number, "Sweet Georgia Brown," and the curtain went down on opening night.

The performance was also noted by Don DeMichael in *Down Beat*.

> Russell was outstanding on "Pee Wee's Blues," during which he worked his way from soft, low-register murmurings to a wailing, squirming set of choruses that brought a number of the listeners to their feet in acclamation.

George Wein had arranged a September tour of Europe for the Newport All Stars—Braff, Pee Wee, Bud Freeman, Wein, Jimmy Woode, bass, and a drummer the band picked up in Germany, Joe Say. The band played concerts and festivals in Berlin, Basel and Copenhagen to sell-out crowds. In Berlin the band played as "Pee Wee Russell's Chicagoans" and, according to the German jazz publication *Jazz Podium*, Pee Wee also proved "his timeless ways as guest soloist with the Thad Jones Quintet which consists of members of the George Russell Sextet."

In Stockholm, the Newport All Stars were on the same bill with Miles Davis, whose drummer, Tony Williams, sat in with Wein's group. The band wound up the tour in London at the Marque Club, where one of their performances was filmed for a "Jazz 625" telecast over the BBC. After that, Wein and company, except for Pee Wee, headed home.

Jeff Atterton had been attempting to get the clarinetist to tour Great Britian for years. With the reaction of the Asian and European fans in his mind and with Mary at his side, Pee Wee appeared as a single in a series of concerts in which he was featured with the top British jazz bands. But just before the tour was to start, his health almost prevented it. The three weeks' travel on the Continent had been grueling. After the last performance at the Marque Club, Pee Wee headed for his hotel, the Strand Palace, complaining of feeling "exhausted." A doctor confined him to his room for three days. But he bounced back again, and on October 15, Pee Wee and Mary were in Manchester. At the insistence of the concert's backer, L. C. Jenkins, general secretary of the Manchester Sports Guild, Pee Wee reluctantly met the press. The following night, he appeared at the Manchester Sports Guild and Social Center with the Alex Welsh band, and on the 17th, he played there with the Johnny Armatage Jump band. One of the highlights of the concert was when two of the finest English clarinetists, Sandy Brown and Archie Semple, teamed up with Pee Wee on a performance of his composition, "Midnight Blue."

The following night, October 18, Pee Wee played at the Manchester Sports Guild with Freddie Randall's band. On the 20th, he played the George and Dragon in Bedford with the Alex Welsh band and on the 22nd he was back at the Manchester venue with the Gary Cox Quartet. The Alex Welsh band accompanied him on the other gigs: October 23 at the Palace Hotel in Southport; October 24 at the Dancing Slipper, Nottingham; October 25 at the Coatham Hotel in Redcar; October 26 at— where else?—Condon's Club in the Red Lion Hotel, Hatfield; October 27 at Crown Hotel, Morden, Surrey; October 28 at Dolphin Hotel, Botley; October 30 at Conway Hall, London; October 31 at Midland Jazz Club, Birmingham. The tour ended on November 1 back at the Manchester Sports Guild.

During the tour, Mary was asked about the type of clarinet and mouthpiece Pee Wee used. The rubber bands were a thing of the past. "He prefers a Buffet clarinet," she said, "and likes a mouthpiece with a fairly open lay. The strength of his reed is $3^{1}/_{2}$ and he isn't fussy about the make. He generally picks one blindfolded or leaves it to me and I adopt the eeny-meenie-miney-mo method," she said.

Back home, Pee Wee found that he had come in third in the *Down Beat* readers' clarinet poll, behind Giuffre and Buddy DeFranco. The jazz fans in America had not caught up with the critics. But Pee Wee did not care. After the frantic activity of the past year, Pee Wee was glad to be home. He spent his time watching television, taking it easy and not playing gigs that he didn't feel like playing. At the age of fifty-nine, he was easing

himself into semi-retirement. Never one to seize opportunities
or to promote himself, his basic reaction to the tumultuous success of the tours in Asia and Europe was that he was glad they
were over and that he had some time to relax. And during the
rest of the 1960s, Pee Wee took life as easy as possible. He
played few jobs outside the jazz concert circuit in the United
States. One special concert close to home took place in the
Museum of Modern Art's Sculpture Garden. The Pee Wee
Russell Quintet, featuring Bobby Hackett, was filmed there by
NBC. A portion of the concert was telecast on the network's
Kaleidoscope program on September 4. A light sprinkling of rain
during the outdoor concert did not dampen the enthusiasm of
the capacity crowd.

On August 13, 1965, Pee Wee played before a crowd esti-
mated at 7,000 at the *Down Beat* Festival at Soldier Field in
Chicago with Newport All Stars Ruby Braff, Bud Freeman,
George Wein, Steve Swallow, bass, and Joe Morello, drums.
Tenorist Stan Getz joined the All Stars for a concluding blues
duet with Pee Wee. The following afternoon in 90-degree heat
Pee Wee played another "Austin High Gang" reunion with
Jimmy McPartland; Floyd O'Brien, trombone; Bud Freeman;
Art Hodes; Jim Lannigan, bass; and George Wettling. Films
were made of the reunion and televised on the local NBC affili-
ate on November 15.

Record producer Bob Thiele, a fan of Pee Wee's from the
early 1940s, began production of a series of jazz albums for a
new label, ABC Paramount, and used Pee Wee frequently over
the next few years. The first album to involve the clarinetist was
with banjoist-singer Clancey Hayes, accompanied by Yank Lawson and His Yankee Clippers. Pee Wee's work on the album was
outstanding. He turned in an especially moving solo on "Dinah,"
during which he apparently detected a flaw and muttered an
audible "aw!" The album garnered four stars in a *Down Beat*
review.

Even with a reduced schedule, Pee Wee's few accomplishments during 1965 were enough to win the *Down Beat* international critics' poll again that year, with 66 votes. Duke Ellington's clarinetist, Jimmy Hamilton, came in second with half that number. Although his standing with the critics was high, Pee Wee chose not to exploit it. He continued to spend most of his time at home watching television and Mary grew more and more irritated with his semi-retirement in front of the tube. One evening, she came home with a large package of artist's materials and dumped it in Pee Wee's lap. "Now paint," commanded Mary. Her whim—intended only to get Pee Wee stirring—provoked a totally unexpected explosion of creative energy.

Pee Wee had never touched a paint brush in his life. A close friend, drummer George Wettling, was an accomplished painter who had studied with Stuart Davis. Pee Wee also knew Davis and many other painters from the days in the thirties at Nick's when the bar was a hangout for artists of all kinds. Many other jazz musicians had also shown strong artistic talent, from Duke Ellington to Paul Desmond, but Pee Wee had never exhibited any inclination.

He unwrapped the package. "What do I have to lose?" he said, remembering the event several months later.

> I took the brush and as a matter of fact, I don't think I even mixed the paint with linseed oil or turpentine or anything. I used the paint straight.

On November 30, 1965, he completed his first canvas.

> After the first picture, I tried a little linseed oil and saw what happened. I found it easier to work with. Well, I figured, what would I paint? A tree or telephone post or something and I said to myself, well, it's only normal I paint something concerning music. Well, what's more musical than a recording date? You think of everything being set up, the boys in the group—and the people in the control room. So I tried to capture on canvas a record date . . . I'm just about as far away from music as I can get,

with the exception of trying to get rhythm into my painting. I
hope there's rhythm in there of some sort . . .

It was, like all of the paintings to follow, a startlingly original
work, as colorful and individualistic as his clarinet playing.
Viewers were struck by the similarity between Pee Wee's
paintings and his music. "He didn't talk about himself or his
life," said Robin Goodman, the daughter of Mary's nephew,
"but he did talk about music. He said he imagined it in his mind
and could see it in colors. When he started painting, I was
reminded of that story because his paintings looked so much like
his music." Pee Wee obviously had been absorbing, perhaps
unconsciously, a great deal about contemporary painting for
many years. Bud Freeman remembered the time in 1950 when
Pee Wee took him to the Cedar Bar and Grill in the Village,
where "well known painters, including Jackson Pollack, Stuart
Davis, and Willem de Kooning, hung out . . . Pee Wee was a
good friend of Stuart and introduced me to him. . . . They
[the painters] were jazz fans."

"Like my playing," Pee Wee said,

> it's a challenge to get in and out of certain choruses. I start with
> certain colors and when it starts getting too dull for me, I say how
> will I get out of this and what color will I use? And one part of the
> picture overshadows the other and it seems to draw too much
> attention to one corner or to the center. If attention is drawn to
> the middle of the painting, then I put something else in the
> middle or the corner to divert the eye. I put a bright red, maybe,
> or a yellow to immediately spread the viewer's vision—in other
> words, what I'm trying to do is not have just one subject stand
> out and the rest as background. I like to have it so that one looks
> at the whole picture and gets a total reaction instead of, shall we
> say, just seeing one object in the center and that's that . . . My
> paintings are not created according to the book—you should do
> this and you should do that. But I'm a rebel. I like to try different
> things.

Pee Wee painted with the canvas board horizontal on his lap or on a chair.

> I didn't know which end of the brush to use. I had a very wild idea of colors. I had no idea of how colors should blend. I had no idea of that, at all. . . . I had visited museums and looked at paintings, of course, but that was the extent of it. I didn't appreciate these things before.

Pauline Rivelli, interviewing Pee Wee about his painting, asked if he was "improvising on canvas." He replied:

> Yes, like playing the clarinet, I've never really aped or copied anybody, I play that way and the same goes for my painting. I'm trying and I'll get there. I may be taking a roundabout way, and it may take time, and I'll make mistakes, but I'll get there somehow. I know my limitations and I know how far I can stretch out.

Rivelli said the bright colors reminded her of the West and Indians.

> I've been told that [said Pee Wee]. And why not? I was raised in Oklahoma and that's Indian territory. At least it was when I was there. I kind of sense the colors from then. The woods, the undergrowth, it all comes back to me. As I kid I had these colors surrounding me. It's very vivid . . . Subconsciously, I guess I'm going back to my years as a boy in Oklahoma.

Within a year, Pee Wee had completed more than sixty bold and powerful abstract canvases. The walls of the apartment were covered with "brightly colored examples of this sudden and prolific burst of creative energy in a new medium," wrote Dan Morgenstern at the time. Mary supplied many of the imaginative titles for the paintings: "Parisian Sewers," "Twins from Mars," "Dance Around the Fire," "The Inner Man" and so on. An experimental film maker, Jud Yalkut, heard about the paintings and decided to make a film of Pee Wee at work with his brushes, but the results of the unfinished project unfortunately seem to be lost. The almost immediate, very positive reaction to his art work helped Pee Wee become more outgoing and confident than he had ever been.

"Now here was a man, intelligent, gracious and well-spoken," said Bud Freeman.

But we never knew it because it was terribly difficult to have any rapport with him because he was so frightened and backward and nervous. You'd get a yes or a no out of him and that was all. And I was not to know how lucid and articulate he could be until he had begun to paint.

One day, I called him to ask him about his painting and he was so lucid and clear, because he'd stopped drinking when he took up painting, that I thought I had the wrong telephone number, I bought his first painting, and he wanted to give it to me, but I bought it for a hundred dollars. Now I could get a few thousand for it.

I acted after a fashion as an agent for him. I'd say that his painting was somewhat after the fashion of Joan Miro. He was so good that people around the world identified more with the idea that he was a famous painter than a clarinet player. So he sold something like 54 paintings for not less than seven hundred dollars a painting. I got three thousand dollars for two of his paintings and he called me up to thank me, and then died four days later. He could have been a rich man just through his paintings. The Australian ambassador wanted to hold an exhibition of his work, a black tie affair in Washington, so I said to him, good God, it'll make you world famous. But he said no, he couldn't go. Why in hell not, I asked him. I don't have a tuxedo, he said. I ranted and screamed at him, but I realized at the end of it all that he didn't think of himself as a painter and he was frightened. So he didn't go. All the paintings were sold, and somebody picked up fifty thousand dollars somewhere.

Meanwhile, the "New Groove" album with Marshall Brown, recorded the previous year, had been issued, and was causing renewed interest in Pee Wee's place in jazz history as well as in the current scene. In a perceptive review of the album, titled "What Ever Happened to the Clarinet?," Martin Williams wrote:

> Today, there are two truly contemporary clarinetists. A reborn Giuffre, now into the most advanced "new thing" jazz, and a reborn Pee Wee Russell, contemporary not only because of his recent work, but because he has been a jazzman bigger than category or style for almost thirty years . . . He is a committed improviser. Although his style has its recognizable characteristics and his use of his instrument its touchingly personal voice, Pee Wee Russell is willing to take chances, to dare, to let the moment find its unique meaning as he plays—indeed it is sometimes as if he were learning to play the clarinet all over again. But improviser or not, it is results that count and Pee Wee Russell is a gifted lyricist . . . it is as if the most recent challenges in jazz had liberated Pee Wee Russell more than ever.

Even though Russell and Giuffre were considered in the forefront of jazz clarinetists at that time, the instrument itself had fallen out of favor. By the mid-1960s, the clarinet was becoming an antique instrument in jazz. It was on the brink of suffering the same fate that befell the banjo and tuba during the transition from classic jazz to swing in the early 30s. While many of the most popular bandleaders during the swing era had been clarinetists—Benny Goodman, Artie Shaw, Woody Herman, Jimmy Dorsey—the tenor and alto saxophone players came into their own during the bebop period beginning in the 1940s. Later, John Coltrane and other modernists brought the soprano saxophone great popularity as well. When Pee Wee was asked about the decline of the clarinet's popularity in jazz he said:

> It didn't progress. In the days of the big bands, the kids adored these men and did their best to imitate them—Tony Scott, Peanuts Hucko, etc. When the big band leaders stopped, so did the kids!
>
> Remember, the bandleaders of the 'thirties and 'forties were glamorous and well publicized and with some old fashioned showmanship "sold" the clarinet to the kids. Guys like Giuffre

and DeFranco have tried hard to gain acceptance for the clarinet
but seemed to have failed. Furthermore, the clarinet is not an
"angry" instrument—it is sweet and gentle—so the avant-garde
looks with disfavor on the clarinet.

I have hope though. I'm optimistic enough to believe that the
clarinet will become popular again—with musicians and fans.
[Then he added, perhaps with his conspiratorial wink], If I *really*
knew the answer, I'd be working.

In another interview around this time, Pee Wee said the clarinet
had been "obsolete now for eight or ten years so it's harder to
fight to make yourself heard. You have to have something altogether different that will cause a little comment." Pee Wee had
enough of that "something altogether different" to win the *Down
Beat* critics' poll again in 1965.

He began 1966 with another Bob Thiele recording session
involving an all-star Ellington aggregation under the leadership
of pianist Earl "Fatha" Hines. Pee Wee fit in as well with the
mainstream swing musicians as he had with Clancy Hayes—or
with Jimmy Giuffre.

In March, 1966, Pee Wee went to Washington, D.C.,
where he appeared for the first time at Blues Alley, with Tommy
Gwaltney's local band. Gwaltney, who played clarinet and vi-
braphone, was the owner of the club and a fervent admirer of
Pee Wee's playing. On some of the numbers, Gwaltney played
clarinet duets with Pee Wee and on others accompanied him on
vibes. Guitarist Steve Jordan also provided strong support. "The
inimitable clarinetist's spirited playing inspired Gwaltney to
some of his finest clarinet playing, and capacity crowds gave the
Gwaltney-Russell clarinet duets great applause," reported *Down
Beat*.

One of the fans who crowded into Jazz Alley was Helen
"Daisy" Decker, Pee Wee's childhood friend who had played
such a vital part in spreading the word about his near-fatal
illness in 1951. Daisy had moved to an apartment in Alexandria,

Virginia, and was a regular at Gwaltney's. Pee Wee and Daisy soon renewed their friendship.

Returning to New York, Pee Wee plunged back into more recording. Another Bob Thiele production again with Yank Lawson's band, took place in May, 1966. The album, "Olé Dixie," received a very generous four stars in a *Down Beat* review. Actually, the album's concept—dixieland warhorses and some contemporary pop standards accompanied by rhythms played with gusto by Chico Hamilton—today sounds very dated. Nevertheless, *Down Beat* singled out "Fidgety Feet" on which "clarinetist Russell, who is especially exciting on this album, seems about to go off like a roman candle in his solo, but he ends it with sparse, descending figures that hint of deep beauty—a thing that Bix Beiderbecke did so well."

Kenny Davern remembered an incident around this time when he asked Pee Wee to substitute for him in the band at the Ferryboat, a club on the New Jersey shore where Davern had been working for a couple of years. Davern was able to get to the club in time for the last set and the band members urged him and Pee Wee to play a duet.

> I was game, so I asked Pee Wee what he would like to play and he said: "You got it, chum." I started to wail away on an E-flat blues [Davern said]. I must have played about eight choruses, and then he started to play. He played so great, and I felt so awful. He couldn't have carved me to pieces more magnificently. So after the gig, I was sitting very morosely as I drove him back. Finally, I said to him: "Did you have to do it to me that bad in front of my friends?" And he replied, "Well, you put me on the line, chum."

It seemed as though the fewer jobs Pee Wee played the more popular he became. He won the *Down Beat* critics' poll again in 1966 and came in second to Buddy DeFranco in the readers' poll. The second album he made with Marshall Brown was issued on Impulse! and again *Down Beat* raved, giving the album five stars.

Russell's playing *is* contemporary, fresh and moving and full of the unexpected, as alive and meaningful today as when he made his initial impact in the late '20's. He has never stood still, has always been a uniquely original player and has always remained himself, defying all attempts at categorization.

The last weekend in May, 1966, Pee Wee appeared at the Atlanta Jazz Festival with the Newport All Stars. "Russell played the blues and 'At the Jazz Band Ball' in optimum spidery style," reported *Down Beat*. He also played a duet with Woody Herman. "I had a ball," Herman said in *Down Beat*. "I want to record with him—just two clarinets and a rhythm section. I'd play melody, and Pee Wee would play around—it would be nice."

Pee Wee played several other concerts during the year, but for the most part he stayed at home and painted. His concern for Mary's health was growing. The doctors she had been seeing insisted that she had an ulcer and that she would have to quit smoking and give up all alcohol, but that seemed to be too much to ask. Mary's nephews, Leo and Sol Goodman, continued to contribute to the couple's finances, supplementing their dwindling income as Pee Wee played fewer and fewer jobs.

One important appearance that year was a concert on October 29, 1966, at the Massachusetts Institute of Technology with another great jazz individualist, trumpeter Henry "Red" Allen, with whom Pee Wee had made the classic Billy Banks and the Rhythmakers sides 34 years earlier. They were backed by a very modern rhythm section of Steve Kuhn, piano, Charlie Haden, bass, and Marty Morell, drums. The concert included Whitney Balliett reading excerpts from his *New Yorker* profiles of Red and Pee Wee.

Allen's career was flourishing in the mid-1960s. Born into a New Orleans musical family, his fiery, unique trumpet style had been serious competition for Louis Armstrong in the late 1920s. In the 1950s, he led various dixieland combinations at brash clubs like the Metropole, often teaming up with Coleman

Hawkins. In the 1960s, however, he was suddenly proclaimed a modernist by trumpet star and bandleader Don Ellis, and his playing received overdue critical acclaim. It was fitting, therefore, that the two aging modernists should include in their concert repertoire compositions by Thelonious Monk ("Blue Monk") and John Lewis ("Two Degrees East, Three Degrees West"), along with two of Pee Wee's blues compositions.

Pee Wee and Allen rode up to Boston together on the train from New York. It was a somber trip. Allen sat quietly, looking down at the floor. Finally, Pee Wee said to him: "OK, Red, tell me what's the matter with you. You know, we used to fight back-to-back in East St. Louis. You can tell me, chum." Allen slowly looked up and told his old friend he had incurable stomach cancer.

The album was the last for Allen, who died a few months later. The concert itself was a "fiasco," according to Balliett. Kresge Auditorium was only half full, and the sparse audience consisted mainly of students apparently attending for extra credit and with no real interest in being there. Nevertheless, the two veterans played at the top of their form. The disc received four stars in a *Down Beat* review by Dan Morgenstern:

> Russell is in fantastic form. He is always himself, but here even more so than usual. His solos are gems, rewarding constant rehearing with new discoveries. This is one of the deepest players in our music, and his conception transcends such essentially meaningless labels as "modern." Eternal would be more like it . . . It had been the fashion to regard Pee Wee's technique as something maverick-like and odd that happens to work for him. But that is foolishness; he knows that horn inside out, and can produce a range and variety of sounds that run the gamut from purity to pure funk.

Pee Wee occasionally sat in with the Jimmy McPartland band during November. McPartland was in residence at The Office, in Freeport, New York, on Long Island, and Pee Wee would show up for the Friday and Saturday sessions.

In February, 1967, Bob Thiele recorded Pee Wee in yet another context. He had the noted arranger and bandleader Oliver Nelson score the backgrounds for a very big band. The album was even less musically successful than "Olé Dixie" had been. The ponderous arrangements sounded, as Kenny Davern put it, "as though they had been written for Woody Herman." General critical reaction to the album was that Nelson's arrangements and the rock-influenced drumming by Grady Tate stifled Pee Wee's lyricism. However, Pee Wee's playing on the album was excellent throughout, strong, eloquent, deeply beautiful. His tone on "The Shadow of Your Smile" was a joy to hear. In the somewhat chaotic double-timed "Pee Wee's Blues," he even managed to reassert some genuine blues feeling.

Thiele's final enterprise with Pee Wee later that year also produced mixed results. The producer assembled another big band, but this one included a small jazz contingent: Jimmy McPartland, Lou McGarity and Pee Wee. The album was a cut-and-splice editing job, aimed at merchandising the old jazz with vocals by Thiele's wife, pop singer Teresa Brewer, and television personality Steve Allen. The band was called "Bob Thiele and His New Happy Times Orchestra," and the jazz sections, sounding as though they had been recorded at a different session and spliced into the big band segments, featured Pee Wee on solo spots on "Charleston," "Betty Co-ed" and two tunes associated with Bix Beiderbecke, "San" and "Changes." It was Pee Wee's last studio recording session.

Outside the recording studio, George Wein continued to provide Pee Wee most of his best jobs. In the spring of 1967, Pee Wee was off on another foreign tour arranged by Wein, this time to Mexico, a country he had last visited when he was a teenager just starting out with the Herbert Berger orchestra. Mary did not accompany him on the trip; her stomach pains had become more frequent and severe. The doctors continued to treat her for a "nervous stomach" and kept her on ulcer medication.

The three-day tour of Mexico, jointly sponsored by American Airlines and the Mexican government, was billed as the "Newport Jazz Festival in Mexico." The package included the Newport All Stars, the Dave Brubeck Quartet, Thelonious Monk and Dizzy Gillespie. The festival started with two concerts in the 2,000-seat Belle Artes Opera House in Mexico City on April 12. The next day, the bands appeared in two concerts at the annual festival in Pueblo, the first time jazz had been included in the festivities, and on April 14, they returned to Mexico City for a concert in the huge, 14,000-seat National Auditorium.

Following the Mexican tour, the All Stars added trombonist Lou McGarity and tenor saxist Buddy Tate, and one of the mainstays at Condon's for so many years, baritone saxist Ernie Caceres. The group played the Wein-organized festival in Longhorn, Texas, April 28 through 30. Trumpeter Kenny Dorham reviewed the event for *Down Beat,* noting that "Pee Wee Russell leads off on clarinet in his very formidable style, which has made him a legend. This ol' cat is very modern and plays some changes that guys of his era would have called avant-garde."

Wein next booked the All Stars to play at the "U.S. National Day," May 25, at the Montreal, Canada, world's fair, Expo 67. The concert, organized by the Institute of Jazz Studies at Rutgers University and produced by Wein, was held at the Place des Nations, an outdoors ceremonial area at the fairgrounds. President Lyndon Johnson attended the National Day ceremonies in the morning; the jazz festival commenced at 5 p.m. with the master of ceremonies, John Hammond, introducing Muddy Waters and His Blues Band. The All Stars, representing "Chicago Jazz," came on next. Other jazz pigeonholes followed: swing represented by Buddy Tate and Roy Eldridge, bebop by Thelonious Monk's Quartet, Afro-Cuban jazz by the Herbie Mann Quintet, and "cool" jazz by the Dave Brubeck Quartet. The weather had been predicted to be sunny, with some cloudy periods. Instead, a bitter cold wind blew and drizzle started just before the All Stars took the stage. The 30,000 fans in the audience bundled themselves up, and a reviewer admitted scurrying for cover from the cold rain and missing part of the concert, but most of the fans stood their ground, feasting on free hot dogs and Pepsi, courtesy of the American taxpayer, and applauding the "American art form" in all of its various manifestations.

Pee Wee hurried back to New York to play a benefit on June 4 at the Riverboat. For once, his heart wasn't in his playing. He had been worried, even frightened, about Mary during the tour with Wein. She was admitted to St. Vincent's Hospital in early May for tests. The work with Wein had helped Pee Wee keep his mind off of Mary's problems to some extent, but returning to New York brought his anxiety to the surface. The test results were worse than anyone had expected: cancer of the pancreas. Incurable, the doctors said; just a matter of a few weeks at most. Pee Wee stayed with her in the hospital room or in the hallway outside. It was a crushing blow.

> I went to see Mary in the hospital [said Phyllis Condon]. I saw her in intensive care. Pee Wee was outside the room crying. He wept as he said she was dying in there. He thanked me for coming. He was very attached to her. I've never seen such agony as Pee Wee went through at St. Vincent's. That scene outside intensive care was something. There was nothing half-way there.

Sol Goodman, Mary's nephew:

> She had been mis-diagnosed. For years, they told her she had ulcers and that her stomach problems were all in her head. They sent her to a psychologist.

In fact, even if the correct diagnosis had been made, there was no cure for cancer of the pancreas at that time. On June 7, Mary died at the age of 56. She was buried in B'Nai Abraham cemetery in Union, New Jersey. Pee Wee was inconsolable.

> It was Mary Russell's intuitive understanding and love that helped to widen the scope of her husband's creative potential [wrote jazz critic George Hoeffer]. If Mary, in her inimitable and

sometimes brusque manner, told the great jazz clarinetist he could accomplish something musically, usually something that he himself was hesitant to try, he could and did with marked success.

Without Mary, Pee Wee's life—even his music—suddenly meant nothing to him. Just when he at last seemed happy and secure, fate made his worst fear a reality: the self-styled loner finally had to face his demons by himself.

13 • • • •

Last Sessions

After the funeral, Pee Wee returned to the apartment. The vibrancy of his brightly colored paintings hung throughout the apartment by Mary, mocked his mood. There would be no more painting. But there would be drinking, and lots of it. He filled a ten-ounce tumbler with vodka and turned on the television set. Human voices took some of the chill out of the room. It had been sixteen years since their reconciliation. Drawing strength from Mary, Pee Wee had progressed from being a physical and mental wreck to having a vitality and confidence he had never exhibited before. With Mary's help, he had been able to achieve and maintain a remarkable level of artistry in his playing. Although he had remained wary of strangers, he had come to speak his mind clearly, and had been reenergized and emboldened artistically. His playing had never received more respect and understanding from fans and critics. But all that had died at St. Vincent's. Without Mary, he felt he couldn't go on. However, there were jobs that he had agreed to do. He needed to fulfill the commitments he had made. He didn't want them to think he couldn't play anymore.

At first, work seemed to help him regain his bearings. His

first job after Mary's death was with Bobby Hackett's band at the Riverboat, a two-week engagement, beginning June 26, 1967, with Ross Tompkins, piano, Bob Dougherty, bass, and Ernie Hackett, Bobby's son, on drums. Roberta Peck, a vocalist, appeared with the band over the July 4th weekend. On June 29, Pee Wee appeared with the Newport All Stars (Ruby Braff, Bud Freeman, George Wein, Jack Lesberg and Don Lamond) in New York. The following day, the All Stars played at the Newport Jazz Festival, and Pee Wee was featured on what one reviewer described as a "soulful, wispy 'Sugar.'" Tenor saxophonist Budd Johnson was added to the group for a couple of numbers. After the festival, the All Stars traveled to Milwaukee for a gig.

Eddie Condon contacted Pee Wee about a job he had lined up for a day in September at Disneyland in Anaheim. It was the eighth annual "Dixieland at Disneyland" show, and Louis Armstrong was to headline with his band. Pee Wee, Yank Lawson, Lou McGarity, Dick Hyman, Condon, Lesberg and Cliff Leeman appeared in "Tomorrowland" on September 30, 1967. A horse-drawn wagon carried the Condon band through the park while other wagons carried bands led by Doc Souchon, Teddy Buckner, the Firehouse Five Plus Two, the Young Men from New Orleans and the South Market Street Jazz Band, down Main Street. But Armstrong's band did not appear: the great trumpet player was ill with pneumonia. "The overlapping sounds as one wagon passed and another approached lent a Charles Ives flavor of massed counterpoint to the parade," reported *Down Beat*. "Seeing Eddie Condon in the futuristic 'Tomorrowland' was a bit incongruous."

On October 8, 1967, back in New York, Pee Wee played a benefit for Sidney DeParis and Hank Duncan at the Village Gate. DeParis died shortly before the benefit was held. Ever since his near fatal illness in 1951, Pee Wee had made a point to play at musicians' benefits.

He next flew to Chicago to appear in a National Educational Television Network series by Art Hodes called "Jazz Alley." Pee

Wee had seen Hodes at Disneyland and had agreed to do the
show then. Hodes recalled Pee Wee in the television studio:

> The set called for a bar and a bartender. No booze, but I'm sure
> the needy didn't come naked. We sailed through the show; I will
> admit a lot of trepidation was running loose. Bits and pieces; like
> Pee Wee looking up and discovering our bartender and putting in
> his order . . . Bob Kaiser (the director) had set us up where he
> thought we should be. Pee Wee was next to the piano and Jimmy
> [McPartland] in the center with the mike just in front of him. As
> I recall, Pee Wee usually leaned against the piano on other gigs.
> But this time he wanted the center spot. He must have asked
> Jimmy several times, because he finally was heard to say,
> "Jimmy, I'll give you five bucks if you change places with me."

In mid-December, Pee Wee opened at Blues Alley in Washington, D.C., with Tommy Gwaltney's band. Tommy said:

> I met Pee Wee at the train station and took him to my house. He
> had reached the point where his playing was not all that great,
> but he was a terrific guy and we became close friends.
>
> He complained of feeling poorly, so Daisy Decker and I ar-
> ranged for him to see Dr. Richard Huffman, an outstanding
> Washington internist. The results of course were disastrous.
> Dick Huffman told me, "This man has absolutely everything
> wrong with him." They were surprised [said Gwaltney], because
> Pee Wee still looked very well.

Despite his physical condition, Pee Wee was still able to con-
vince his listeners that all was well. John Segraves reviewed Pee
Wee's performance at Blues Alley for the *Washington Star:* "Age
is not bothering Pee Wee one bit. The notes, clear, crisp and
warm, are still flowing in torrents. See for yourself."

Pee Wee's friend, critic George Hoefer, had talked to him at
the beginning of 1968 about writing his biography. And Jackie
Gleason, a devoted fan of Pee Wee's since the years at Nick's
when Gleason was a struggling burlesque comedian, had talked
about the possibility of making a film about Pee Wee's life with
Jimmy Stewart in the lead role. Pee Wee had even agreed to

cooperate with Hoefer—reluctantly—and the writer had started researching Pee Wee's career, going through hundreds of issues of *Down Beat* magazine, but before the project could get off the ground, Hoefer died.

"Pee Wee was distressed at the death of George Hoefer," said record producer Hank O'Neil:

> Pee Wee asked Eddie Condon if he knew anyone who might be able to pick up the pieces. I was Ed's choice.
>
> I didn't know, when I entered his sparsely furnished project-type apartment, that he was still devastated over the loss of his wife, was rarely playing and had simply given up. All I saw was a man whose music I knew to be glorious but who seemed to be the saddest, most quiet and withdrawn person I'd ever encountered.
>
> When I saw him in February, 1968, I was shocked. I was in Washington, D.C., on business and one night Squirrel Ashcraft and I went to Blues Alley to see Tommy Gwaltney. Pee Wee and some friends were two tables away and we visited for a few moments. A few months earlier he just evidenced an awful depression; now he was not only depressed but looked dreadful as well.

Pee Wee spent a lot of his time in the Washington area, living at Daisy Decker's apartment. "Somewhere along the line," said Kenny Davern, "the whole new life he planned fell through. He was torn between the possibility of a new life with Daisy—he told me he was going to marry her—and feeling like a traitor to Mary. He stopped painting after her death and felt very worthless and useless. He was very self-deprecating. Helen was a bright spot for a while, but she was also part of a group that drank a hell of a lot."

After completing the Blues Alley engagement, Pee Wee returned to New York and stayed in his apartment. He was drinking vodka incessantly, rushing headlong into oblivion. Mary's nephews, Sol and Leo Goodman, continued to provide financial support during this period. Pee Wee spent his time watching "my westerns and detectives" on televison.

Davern and his wife Elsa tried to look after Pee Wee, and on occasion invited him to dinner and then suggested he spend the night. One night, while he was staying at their apartment, they were awakened at 4 a.m. by strange noises.

> I heard some thumping and went to the door and opened it [said Kenny Davern]. There was Pee Wee standing with a bottle of Smirnoff in one hand and a bottle of Maalox in the other. I asked him what was the matter, and he said, "I can't sleep, chum." He came into the bedroom started smoking cigarettes and talking for a while. Finally he tells us what the problem is: he can't decide whether to drink the vodka first and pour the Maalox on top of it, or drink the Maalox first to coat his stomach and then drink the vodka. It was a huge dilemma for him.

Once after Davern had pulled up outside 28th and Eighth to let Pee Wee out, Pee Wee grabbed for the keys and turned the car off. "Come on, chum," he said, "I want you to have a painting. Now come and get it." As they entered the apartment, Davern saw all of Pee Wee's paintings—about 90 of them—stacked up four or five deep against the walls of the foyer. The walls themselves were bare. Beyond the foyer was the living room with Pee Wee and Mary's recliners, the television set, a pole lamp and an air conditioner. Davern said

> I expected him to just grab one and say "here," but instead he told me to pick one out. He sat down in a chair and I went through every one of them and pulled out two I really liked. One was "Belly Dancer," which was one of the few representational paintings and the other was called "Faces in the Crowd."
>
> After about an hour of contemplation, with no help from him, I pointed to one and said that it was the one—"Faces in the Crowd." He looked at me and exclaimed, "You would! That's my pet!" So, I said, "Keep it, then." "No, no, you got it. But before you take it, just let me see it."
>
> He took the painting and brought it up to the pole lamp. He placed the whole canvas right into his stomach and held it flat.

With one hand he went across the entire surface of it, as though he were feeling the typography, the texture.

The painting contains five faces, the center one of which, as Elsa Davern was quick to point out, is of Mary, and one on the left, with what looks like patches on his face, is Pee Wee.

The first known job he had in 1968 was a concert given at Kenny's Steak House by the New York Hot Jazz Society. Surrounded by old band mates like Ruby Braff, and with tenor sax star Zoot Sims, Pee Wee seemed to blossom yet again. "All Russell's solos were notable," wrote Whitney Balliett:

> In "Ain't Misbehavin," he played a high, inquiring statement as if he were trying to wring the last melodic secrets from that cherished tune. In "I Want to Be Happy," he slid along just beyond the edge of tonality, and in a slow blues he delivered a first chorus that, in its jumble of rare notes and backings and fillings, teetered wildly at the edge of the abyss. He never fell in. . . . Russell, God love him, is back.

Pee Wee's next job found him in New Orleans at "Jazzfest '68" on May 16, playing on a Mississippi riverboat, the S.S. *President*, with Captain Verne Streckfus at the helm. The cruise featured three bands. The "New Orleans" group was under the leadership of Sharkey Bonano, the leader of the other band at Nick's when Pee Wee first started working there. The "Chicago" band was led by Art Hodes and the "New York" band by Max Kaminsky. Strangely enough, for once Pee Wee was not miscast as a Chicagoan, appearing instead in the New York contingent, with Kaminsky, Bob Haggart, Senior and Junior on bass and drums respectively, Herb Gardner, trombone, and Dick Hyman, piano. The band started off with a hard-hitting "Dippermouth Blues," with Kaminsky playing a fiery solo and Pee Wee sounding very inventive and self-assured.

The crowd, described by one of the audience as "2,400 noisy bodies," however, was in an uproar of New Orleans-style party-

ing. Everyone was shouting over the music, which effectively
was drowned out. When a tune ended, Kaminsky tried to quiet
the merry makers. "You gotta get quieter," he yelled. The volume increased. Finally, Kaminsky muttered, "Well, OK," ad-
mitting defeat. No one had heard him.

They next played "Squeeze Me," more for themselves than
the audience. Pee Wee played a brilliant second chorus, unheard by the audience. Kaminksy introduced the Haggarts' relatively soft bass-drum feature, "Big Noise from Winnetka," and
tried again. "You gotta be a little more quiet please!" he yelled.
When the volume increased in response, he screamed, "Shad-
dup!" A few people applauded. Dick Hyman's fleet rendition of
"Maple Leaf Rag" was completely ignored, and the full band
returned for a perfunctory "Muskrat Ramble." They left the
stand virtually unnoticed. It had been a humiliating evening.
"The audience was wildly responsive," reported Charles Suhor
in *Down Beat,* "and soon after Kaminsky opened his set with
'Dippermouth Blues,' dancing broke out. Kaminsky earnestly
urged the crowd to quiet down, unaware of the fact that when a
New Orleans audience shouts and dances and spills beer on the
floor, the band is winning, not losing."

Pee Wee next appeared with Kaminsky at the relatively sedate Ken Club in Boston on June 4. He was featured the following month at Newport with the Alex Welsh band, the British
group that had backed him on his tour of England four years
earlier. He was still able to cast his spell over the audience with
a heart-wrenching performance of "Pee Wee's Blues." On the
few occasions when he appeared in public, Pee Wee looked as
impeccable as ever: his thinning, graying hair combed neatly
back, his clothes well pressed and his shoes shined to a high
gloss. He looked every inch the successful, urbane New Yorker.
But those who saw him off the bandstand said that he was often
confused and very depressed. Suddenly, for no apparent reason,
he would burst into tears.

From October 21 through October 26, 1968, he played at

George's Spaghetti House in Toronto with the Art Ayre trio. Ayre, who usually played the organ and had made two records with flutist Moe Koffman on that instrument, switched to piano for the engagement. Ron Parks, an electric bass player, and Jerry Fuller, a drummer, completed the group. The trio usually played the current pop tunes, but for Pee Wee they attempted some standards such as "Paper Moon," "Talk of the Town" and "Lazy River."

A moving article in the *Toronto Star,* October 26, 1968, by Jack Batten, painted a poignant portrait of Pee Wee:

> Pee Wee Russell is 62 years old and for nearly 40 years he has been a travelling jazzman, moving around the world, playing his clarinet with every marvelous jazz musician from Bix Beiderbecke to Thelonious Monk. He may, all in all, have established himself as the best, most unique clarinetist in jazz, but that distinction has hardly blessed him with riches, and this week, as in countless other weeks, he found himself on the road, playing at George's Spaghetti House and living out of a tiny room, number 1168, in the King Edward Hotel.
>
> When I went around to call on him one afternoon early in the week, he was fumbling with the top drawer of his bureau. He slid it open and a bottle of Smirnoff toppled off the Gideon Bible inside. He tossed back his arms, imitating chagrin the way W. C. Fields used to do it, and gave the bottle a long, malevolent look. It was three-quarters empty.
>
> "Well, well, well, who's been into here, I wonder?" he said, mildly befuddled. "Mmmmmmm, must have been me. Nobody else living in the room."
>
> He poured three fingers of the vodka into an empty glass—no ice, no tonic, no juice—and drank it empty in one smooth swallow. He chased it with some milk taken straight from a carton, then maneuvered his way to the bathroom and mixed a dose of bicarbonate of soda. He belched a couple of times and excused himself.
>
> A copy of his most recent album, "the Spirit of '67," was sitting on the bed in the room, and Russell looked carefully at the

record's jacket. It was decorated with a reproduction of one of his own paintings, a neatly organized semi-abstract, swarming with bright, happy colors, a little in the style of Fernand Léger.

"I started paintin' just all of a sudden back in '57, '58," he explained. "It was my wife's idea and maybe I did all right. I had a show of a bunch of my pictures at Rutgers University and I sold one to a collector in Australia for $800. The price was his idea. I thought it was ridiculous."

"But I'll tell you one thing—my wife and another fella once took me on a tour around all the art galleries in New York, and when we got back I thought about all the different painters I'd seen that day and I began thinkin' about the way they did certain things and about how maybe I could do things their way. Then I stopped thinkin'. I said to my wife, "I'm not going in another goddamed art gallery for the rest of my life. I'd end up paintin' like somebody who isn't me!" I gotta do it my own way. And, I'm telling you, it's the same playing the clarinet—I gotta do it my own way.

"I like playing, I like keepin' at it. These fellas who think they've got it made, they're through the moment they think that, you know. I don't think I got anything made, not my playin' anyway. I always think I'm going to get better."

He took another long look at the Smirnoff bottle, which was empty, and I left him there in his hotel room

"George's was in the red light district of Toronto," recalled jazz researcher and writer Joe Showler. "The joint sold pizzas. Pee Wee looked very lonely and lost, like 'What am I doing here?' He didn't seem to care if he lived or he died. He complained about the band. 'These guys don't know what they're doing,' he said. They didn't know any of his tunes. The place was empty—only about 20 people in there." Pee Wee complained to Showler about feeling ill, that he felt he should be in a hospital. But when Showler offered to drive him to the hospital, he refused.

Returning to New York on October 25, he appeared one last time at Town Hall, the scene of some great triumphs, in a Hot

Jazz Society concert that was produced by Frank Driggs and Jack Bradley.

Ironically, Pee Wee won the 1968 *Down Beat* International jazz critics' poll and, for the first time since 1944, he also won the readers' poll. The fans had finally caught up with the critics—too late.

Pee Wee's next job—his last—took him to Washington, D.C., on January 20, 1969, to play with George Wein's Newport All Stars at the Sheraton Park Hotel for one of Richard M. Nixon's many inaugural balls. Pee Wee's health was precarious, but that night he seemed to be his old self. "He had tremendous recuperative powers," Wein said, recalling how well he played that night.

After the gig, Pee Wee returned to Daisy Decker's apartment, where he was staying. He was next scheduled to sit in with Tommy Gwaltney's band at Blues Alley during the first week of February, but he told Gwaltney that he didn't feel well enough to play. "He frequently came down to Washington just on weekends," Gwaltney remembered. "Everybody dug Pee Wee and he liked it here. He was seriously considering moving to Washington. But he felt poorly. I didn't ask him to play because I knew he didn't feel like it. He and Buck Clayton chatted for a while, then Pee Wee and I went upstairs to the office. We were alone.

P.W. "I'm sick, I'm in bad shape. The booze, my nerves. Everything."

T.G. "Why don't you go to the hospital and get straight?"

P.W. "I can't. I can't get in a hospital in New York."

T.G. "Oh, come on."

P.W. "No, I can't get in." He was emphatic.

T.G. "Want me to try to get you in here?"

P.W. "Get me in anywhere. Somewhere. Sooner the better."

The next morning, February 9, 1969, a telephone call from Dr. Bill Young woke up Gwaltney: "I have a bed for Pee Wee at

Alexandria Hospital. Get him over here this morning and have
him admitted as soon as you can." Gwaltney called Pee Wee.

"Get ready. They'll take you in Alexandria Hospital now."
"Now?"
"Now. I'll pick you up in 20 minutes."
"Wait. Wait a minute. The cocktail party for you guys. Can't
I wait and go to the hospital after the party?" [A few close friends
were holding a small afternoon affair for Gwaltney and his new
wife, Susie, which Pee Wee expected to attend.]

Gwaltney persisted. "The man says come now. Remember
what you said last night. 'The sooner the better.'" There was a
long pause and then Pee Wee's voice, very softly, said, "Okay,
come on, Tommy."

Pee Wee's bag was packed when Gwaltney arrived at the hotel.

He was nervous. Very apprehensive. He'd gone through a pack of
those skinny Benson and Hedges cigarettes in about a half an
hour. He asked me to telephone Lee Goodman in Union, New
Jersey. I did, and they talked for a few minutes. Pee Wee held
his hand over the receiver and said to me, "How much money
will I need?"

"Do you have any hospitalization?"
"I think so. Blue Cross, Blue Shield, some damn thing."
"You won't need any money."
"I don't know. How much is this?"
I counted it. $165. "You don't need any money." He told Lee
Goodman to send him some money.

"Can I have a drink before I go?"
"You can have two drinks—three drinks—five drinks—but
they're cutting you off cold-turkey at the hospital so this will be
it."

Pee Wee was really nervous. I mean even for him. He had
two more. His face got pink and the conversation during the ten-
mile drive from Washington to Alexandria was animated. "I've
got to get well. I've got three dates with George Wein." He pulled
a sheet of paper out of his pocket and read to me "N.Y.U. and,
what the hell is this? Fordham, then, well I can't make this out,

but I've got to make these dates." The dates were several months off.

"You'll make them."

Finally admitted. In pajamas. Sitting on the edge of a hospital bed in his small but private room. Smoking still another skinny cigarette, hands shaking. They had found the half-pint of vodka he had carefully stashed. Pee Wee was defenseless.

"I visited him once or twice each day," Gwaltney said.

Only three visitors were allowed: my wife, Helen Decker and I. The visits were short, almost meaningless. He was under heavy sedation and very drowsy. He talked more at first, but gradually, as the days slipped by, it became harder to understand him. The doctors seemed satisfied at first; then concerned about his lack of response to treatment. No progress.

Pee Wee didn't complain much. He said his stomach was bothering him some but dammit he was hungry. "Get me something to eat." The hospital authorities said indeed Mr. Russell has a good appetite. But he kind of lost interest in the whole thing. I saw him every day. Friday was not so good. I understood little of what he said. Something about Danny Alvin—getting an apartment with Danny Alvin.

I finally started to leave and walked out of his room into the hall. He started raising hell. A young hospital attendant, Vernon Simmons, who had taken a liking to Pee Wee, went in and came right back out—"Mr. Russell wants to see you."

Pee Wee paused just for a couple of seconds and said very clearly, "Thank you. Thank you for everything."

I patted his shoulder. "Okay, pal, I'll see you later."

On Saturday morning, February 15, Dr. Young called Gwaltney to tell him that Pee Wee had died at 5:30 that morning. He was a month and a half short of his 63rd birthday. The death certificate gave the immediate cause of death as cerebral edema due to acute and chronic pancreatitis. The other significant condition contributing to death was Laennecs Cirrhosis of the liver. (Dr. Ben C. Jones was the attending physician at his death.)

Daisy Decker wrote in the *Arlington Journal* that "a good man and a great genius, Charles Ellsworth 'Pee Wee' Russell died quietly in his sleep . . . He wouldn't let Tommy or me tell anyone that he was sick. 'They might think I couldn't play anymore,' he said." When word of his death reached New York, the jazz world there was shocked. To his friends he had not appeared to be any worse than usual when he set off for Washington. Rumors spread that he had died at the hands of incompetents, that the doctors had not known that he was an alcoholic and injected him with medication that was sure to kill anyone in that condition, that he had died of delirium tremens. But there is no evidence to support those claims. Throughout his life, he had been afflicted with physical ailments: an inability to eat solid food, an addiction to alcohol, and acute nervousness and anxiety. A "good day" for Pee Wee was when his only ailment was a blinding hangover. That he had lived to the age of sixty-two in any event was miraculous.

Lee Goodman made the funeral arrangements, and as Pee Wee had requested, he was buried next to Mary. The service was held in Maplewood, a small town in northern New Jersey. Rabbi Barry H. Greene of Congregation B'Nai Jeshurun of Short Hills, delivered the eulogy, saying the "angular sounds from his clarinet" could be likened to the "inspirational highlights" of Biblical psalms.

Nat Pierce, Johnny Windhurst, Ruby Braff, Gene Gardella, Lee Goodman and Jeff Atterton served as the pallbearers. Louis Armstrong, Louis Prima, Joe Sullivan and Jimmy Giuffre were among those who sent wreaths. Giuffre's card read, "Keep Swinging." More than forty jazzmen attended the funeral, including clarinetist Joe Murayni, Kenny Davern, Marian and Jimmy McPartland, Buzzy Drootin, Zutty Singleton, Bobby Hackett, Vernon Brown, Lee Wiley, Yank Lawson, Jack Palmer, Jack Lesberg, Johnny Blowers, Marshall Brown, George Wein, critic Charles Edward Smith and Eddie Condon.

"The only time I saw Eddie cry in all our years together,"

said Phyllis Condon, "was at Pee Wee's grave during the final moments of the ceremony. He sobbed out loud. I was shocked."

Many of the musicians went to Lee and Estelle Goodman's home after the services. Lee Wiley sang several songs, accompanied by Jack Palmer on piano, and Condon's sobs. J. C. Higginbotham, who had been drinking heavily, left only after he was bribed with a bottle of Scotch. Lee Goodman presented Kenny Davern with Pee Wee's two Buffet clarinets (serial numbers 73114 and 26257). Davern lent one of them to Eddie Cook, publisher of *Jazz Journal International,* and gave the other to the Institute of Jazz Studies at Rutgers University. Davern still treasures the rubber Conn that he and Pee Wee found under the recliner.

The gravestone was inscribed Charles "Pee Wee" Russell, A Beloved Husband, and gave 1907—one year late—as his birth date. Eulogies, many with numerous biographical inaccuracies, appeared in the jazz press throughout the world. A few concentrated on his eccentricities (one was headlined: "Death of an Odd-Ball," by Benny Green), but most were dignified memorials to a man whose music could not be forgotten. A tribute was printed in the *Congressional Record* of February 27, 1969, by Congressman Seymour Halpern of New York. And Pee Wee was inducted into the *Down Beat* Hall of Fame in 1969.

One of the most eloquent obituaries in the jazz press was written for *Down Beat* by pianist Dick Wellstood, who had played with Pee Wee a few months earlier at a Town Hall concert.

> Pee Wee Russell is dead. A simple fact that can be accepted, abstractly, a piece of intellectual exercise like memorizing a solution to a chess problem. And then we re-arrange the pieces and set up another puzzle. Or would like to. But this time we can't. The miracle of Pee Wee's playing is dead, that crabbed, choked, knotted tangle of squawks with which he could create such woodsy freedom, such an enormous roomy private universe . . . Pee Wee is dead. A simple fact. I pick it up, bang it on the

floor and put it in my mouth like a child. And spit it out uncom-
prehended.

Bobby Hackett wrote:

> Pee Wee and I were very close friends for many years and what
> little musical knowledge I may have I owe plenty to him. He was
> truly a great artist and a very honorable man. His music will live
> forever, along with his wonderful spirit. I'm sure we all miss
> him, but thank God he was here.

Whitney Balliett, always one of Pee Wee's most understanding
supporters, wrote in the *New Yorker:*

> Most lives are uphill but Russell's was almost perpendicular. He
> was a shy man trapped in a sausage body and a clown's face. He
> sidled rather than walked and his deep voice seemed to back out
> of him, producing long, subterranean utterances that were ei-
> ther unintelligible or very funny. And when he tried to escape
> into alcohol he became mortally trapped. His style—the knotted,
> chalumeau phrases, the crazy-legs leaps over the abyss, the
> unique *why?* tone, the shining use of notes that less imaginative
> musicians had discarded as untoward—was, paradoxically his
> final snare and his glory. People laughed at it. It was considered
> eccentric, and because eccentricity, the kindest form of defi-
> ance, baffles people, they laugh. But those who didn't laugh
> understood that Russell, behind his Surrealistic front, had dis-
> covered some of the secrets of life and that his improvisations
> were generally successful attempts to tell those secrets in a new,
> funny, gentle way.

As he was leaving the funeral service, New Jersey jazz enthusi-
ast Jack Stine met Lee Goodman:

> Lee mentioned in passing that Pee Wee had often expressed a
> sorrow that, childless, he had not had the opportunity to give
> some deserving kid a leg up, so to speak, in a jazz career. The
> remark impressed me, and on the way home I decided to see if
> there was something I could do to implement this wish.
> I called Lee back and asked if I could have his permission to

run a small benefit for Pee Wee, the idea being to raise perhaps a couple hundred dollars for Lee to award to a youngster graduating from some local high school. Lee was enthusiastic and grateful and so I started out. Once my intentions were known, I began to get calls from musicians like Marian McPartland, Eddie Condon and many others, to see if they could help—and it soon was apparent that I had a tiger by the tail.

The first year's concert raised around $3,000, and we have been holding these concerts every year since.

Out of the "Pee Wee Russell Memorial Stomp" grew the New Jersey Jazz Society, now numbering more than a thousand active members. The first Stomp was held Sunday, February 15, 1970, in the Martinsville Inn, and featured music by Chuck Slate's traditional jazz band and guests. On display were twenty-four of Pee Wee's paintings. Helping to make the event successful were frequent radio plugs by Arthur Godfrey and others. The Stomp continues to be an annual event. The money raised is still awarded to deserving high school students. Among recipients of the award have been Terence Blanchard, trumpet, and Harry Allen, tenor.

Interest in another of Pee Wee's artistic expressions, his painting, also lives on in the homes of the fortunate few who possess them. Rather than regarding the paintings as curiosities that remind one of Pee Wee Russell, most owners of Russell's works express genuine fondness of them and proudly display them. An exhibition of Russell's paintings was mounted in 1978 at the Hunterdon Art Center in Clinton, New Jersey. In 1986, another exhibition was held at Sordoni Art Gallery at Wilkes College, Wilkes-Barre, Pennsylvania. The Institute of Jazz Studies at Rutgers University in Newark, is the repository for the bulk of the paintings.

Pee Wee's unique creations in music, of course, are his claim to immortality. "There are still the recordings," wrote bassist Bill Crow, "reminding us of how wonderfully Pee Wee's playing teetered at the edge of musical disaster, where he struggled mightily, and prevailed." The influence of his style lives on in the work of such contemporary masters as Kenny Davern, Bobby Gordon, Frank Powers and Frank Chace, among many others. Pee Wee's haunting composition, "Pee Wee's Blues," continues to be recorded with some frequency. Within recent years Davern, Powers, Gordon and west coast clarinetist Ham Carson have all released their versions of it; Joe Muranyi featured it nightly while he was with Louis Armstrong's All-Stars.

But as the merchandising of jazz becomes more and more limited to the "big names" the major record companies have promoted over the years—Duke Ellington, Louis Armstrong, Billie Holiday, Benny Goodman, Miles Davis, Thelonious Monk, Dave Brubeck—great work by others, especially those musicians who, like Pee Wee, were nearly always sidemen, has been neglected. Nevertheless, many of Pee Wee's finest recordings can still be found. An uncompromising artist who shunned the commercial to create the beautiful, Pee Wee always had a loyal core of fans and critics who understood what he was doing and appreciated his creativity. Although his understanding of his relationship with the rest of the world was often confused and twisted, he was able to create so much for so many years and at such a remarkably high level, a testament to his basic artistic nature and his compulsion to improvise new stories night after night. As critic George Frazier put it, "Pee Wee *is* jazz."

A Selected Discography

Although Pee Wee Russell was regarded by many to be an "uncommercial" musician, his music was frequently preserved on record. Fortunately for us, every facet of his jazz career was well documented. Almost all of the discs he made during the 78 rpm era were reissued on LPs and many are now finding their way onto compact discs. Examples of his work with Red Nichols' groups are available on an English release, ASV Living Era CD AJA 5025R (Red Nichols and His Five Pennies: Rhythm of the Day); on a Swedish compact disc, Tax S-5-2 (Red Nichols, 1929), which also includes the three outstanding titles with the Louisiana Rhythm Kings, and on Bluebird 3136-2-RB (The Jazz Age: New York in the Twenties). The few sides Pee Wee recorded with Bix Beiderbecke are all available on several CDs, but the release on Columbia CK 46175 (Bix Beiderbecke Vol. 2) of "Crying All Day" is probably the most easily obtainable. All of Beiderbecke's recordings were issued on a 9-CD boxed set of Italian manufacture, IRD BIX 4. "Hello Lola" and "One Hour," Pee Wee's seminal recordings with Coleman Hawkins and the Mound City Blue Blowers, are on Bluebird 9683-2-RB (Three Great Swing Saxophones). ASV Living Era CD AJA 5040 R (Jack Teagarden: I Gotta Right To Sing the Blues) contains two of the titles from the Jack Teagarden session with Fats Waller which produced "That's What I Like About You" and "You Rascal You." The former title is also on Columbia CK 40847 (The 1930s—The Jazz Singers).

Surprisingly, thus far only a few odd tracks are available on compact disc of sessions with Red Allen, issued as Billy Banks'

Rhythmakers, and of Pee Wee's work with Louis Prima. On the other hand, an originally rejected Teddy Wilson session which paired Pee Wee with Hot Lips Page has been released on Classics 548 (Teddy Wilson, 1937–38).

During the last years of the LP era in the late 1980s, Mosaic Records released three boxed sets containing all takes of every Commodore session for a total of 66 discs, many of which contained previously unissued material by Pee Wee. While all of Pee Wee's Commodores are not yet available on compact disc, there is a good sampling of his work from that period on Commodore 7007 (Jammin' at Commodore), with the first sessions, including "Serenade to a Shylock"; 7009 (Jazz in New York— 1944), which features his work with Bobby Hackett, Miff Mole, and Muggsy Spanier; 7011 (Jazz A-Plenty), which includes "That's A Plenty" and "Panama" recordings with Wild Bill Davison; and 7015 (Ballin' the Jack), which features the sides with Fats Waller and Pee Wee's famous recording of "I Ain't Gonna Give Nobody None of My Jelly Roll." Eddie Condon's "Town Hall" concert broadcasts—beginning in May 1944 and continuing through April 1945—are being issued in their entirety (including the still-spry narration) on a series of 2-CD sets from Jazzology. One of the concerts, which preceded the start of the Blue Network broadcasts covered by the Jazzology releases, has been issued on Jass J-CD-634 (Eddie Condon Live at Town Hall). The bulk of two Condon Associated transcriptions sessions from this same period are on Stash ST-CD-530 (The Definitive Eddie Condon, Vol. 1), which also includes alternate and incomplete takes.

The half-dozen titles produced at the first recording session to be issued under Pee Wee's name, recorded August 31, 1938, have been reissued on Original Jazz Classics OJCCD-1708-2 (Jack Teagarden and Pee Wee Russell), although, despite ample time available, the CD omits the two alternate takes. A sampling of his work with Bobby Hackett's band is currently on CBS Portrait RK 44071 (Bobby Hackett: That Da Da Strain), but the sound quality of this release is poor and it has only sixteen tracks, omitting several important sides. Bud Freeman's Summa Cum Laude band is well represented on Affinity AFS 1008, and the first Bluebird session by the band is also on Bluebird 6752-2-RB (At the Jazz Band Ball: Chicago/New York Dixieland). The lovely session with Lee Wiley (Fats Waller on piano) is on Audiophile ACD-1 (Lee Wiley Sings Porter & Gershwin), and the four sides cut with Cliff Jackson's Quartet are on Pickwick PJFD 15001 (Black and White and Reeds All Over).

A burst of recording activity followed Pee Wee's recovery from a near-fatal illness in 1951. Two of the best sessions with George Wein groups are documented on Black Lion BLCD 760 909 (Pee Wee Russell: We're in the Money).

Pee Wee's work with Ruby Braff can be sampled on Bluebird 6456-2-RB (Ruby Braff—This Is My Lucky Day). A superb 1957 session by Pee Wee with a rhythm section is out on Fresh Sound FSR-CD 126 (Portrait of Pee Wee), which also includes his February 1958 session with Ruby Braff and Vic Dickenson. The latter session is also on DCC DJZ-611 in stereo (also titled Portrait of Pee Wee, after the original LP). "Pee Wee's Blues" came from this session and remains one of the finest performances of it. His appearance on the television program "The Sound of Jazz" is available on several video tape issues and on Pumpkin 116 (an LP issue only), and the rehearsal session can be found on Columbia CK 45234 (The Sound of Jazz). Pee Wee's "modernist" career took off with this telecast and his duet with Jimmy Giuffre still remains a fascinating lesson in jazz.

Many of the recordings he made during the last decade of his life demonstrate his renewed energy and creativity. The 1961 date with Coleman Hawkins can be heard on Candid CCD 79020 (Pee Wee Russell/Coleman Hawkins—Jazz Reunion). Pee Wee's 1963 Newport Festival appearance with Thelonious Monk is on CBS Sony 30DP-522. Both 1963 Monterey Jazz Festival sets by the Jack Teagarden All Stars are on Grudge 4523-2-F. A highlight is the inspired blues duet between Pee

Wee and Gerry Mulligan. Finally, the Tokyo concert from Pee Wee's triumphant tour with Eddie Condon's All Stars is captured on Chiaroscuro CR(D) 154 (Eddie Condon in Japan).

In addition to his appearance on the Sound of Jazz, Pee Wee can be seen on another VHS video tape release: "America's Music: Chicago and All That Jazz," Vintage Jazz Classics VJC-2002. While some of Pee Wee's important recordings have not yet been reissued on compact disc (for example, none of the Marshall Brown material is currently available), we can be certain that his unique sound and style will continue to fascinate listeners as long as there is an audience for jazz.

For a complete discography listing all issues on all formats of more than 340 sessions by Pee Wee Russell, including broadcasts, concerts, private recordings, and films, see *Pee Wee Speaks: A Discography of Pee Wee Russell,* by Robert Hilbert, published by Scarecrow Press.